PLUS 5 NEW MINIWEAPONS

MINI
WEAPONS
OF MASS DESTRUCTION™
TARGETS

100+ TEAR-OUT TARGETS

JOHN AUSTIN

CHICAGO REVIEW PRESS

To the junior sharpshooter in training, may you handle this marksmanship manual with the discipline and professionalism exhibited by the real men and women who serve and have served.

Your move, partner!

Join the MiniWeapons army on Facebook:
MiniWeapons of Mass Destruction: Homemade Weapons Page

Cover and interior design: Jonathan Hahn
Illustrations: Austin Design, Inc.

MiniWeapons of Mass Destruction is a trademark of Austin Design, Inc.
All rights reserved

© 2012 by Austin Design, Inc.
All rights reserved
Published by Chicago Review Press, Incorporated
814 North Franklin Street
Chicago, Illinois 60610
ISBN 978-1-61374-013-2
Printed in the United States of America
10 9 8 7 6 5 4 3 2 1

CONTENTS

INTRODUCTION

It's time to master your homemade arsenal with *MiniWeapons of Mass Destruction Targets*. The sole purpose of this guide is to improve your accuracy and precision, training you to shoot your Mini-Weapons safely and competitively.

Designed to complement the MiniWeapons building guides, this book contains a variety of large tear-out targets and mini tabletop targets to shoot at. They include various skill challenges, battle scenarios, and mission-specific training exercises. On the back of each target are official scoring rules for both basic play and marksman-level competition. To set up the targets, see the section on creating a controlled shooting range (page vii).

If you don't yet have one of the building guides, don't worry. *MiniWeapons of Mass Destruction Targets* also shows you how to build a small selection of bonus MiniWeapons to get started! Each of these target shooters is constructed from everyday items, costs only pennies to build, and takes just a few minutes to construct. And the targets in this guide aren't just for homemade weaponry; they can also be used with Airsoft, paintball, and BB guns, and any other projectile launcher you may have.

This is a book for sharpshooters of all ages. It will inspire creativity, encourage experimentation, and fuel the imagination. Some of the MiniWeapons are great representations of real-life counterparts used by competitive shooters, making this an excellent way to hone your skills in a controlled environment.

Finally, this book is for entertainment purposes only. Please review the safety instructions (page vi) for your personal protection. Build and use these projects at your own risk.

PLAY IT SAFE

The unexpected can always happen! When building and firing MiniWeapons, be responsible and take every safety precaution. Switching materials, substituting ammunition, assembling improperly, mishandling, targeting inaccurately, and misfiring could all cause harm. *Eye protection is a must* if you choose to experiment with any of these projects.

Always be aware of your environment, including spectators, and keep any projectile launcher pointed in a safe direction. *Never point a launcher at people, animals, or anything of value.* Some projects in this book use darts, which have dangerous points and can easily harm or damage whatever they come into contact with. Shoot or throw the darts only at a controlled target—away from spectators. Elastic and latex shooters fire projectiles at unbelievable force and can also damage unintended targets. Small BB or spitball ammunition can cause serious eye injury. In fact, ammo, no matter what the material, can cause harm. It is important to remember that since miniweaponry is homebuilt, it is not always accurate. Ammo can and does fly off-target.

Weapons—especially handgun-inspired designs—should never be painted to look realistic; instead, choose a bright color (orange, red, yellow). *Never* take or transport any of these projects on public transportation, such as an airplane, bus, or train. If you are a student, note that the projects in this book could violate your school's "no weapons" policy. If you bring these projects onto school grounds, you could be subject to severe punishment. These projects are to be used at home.

Some of the projects outlined in this book require various tools, such as hobby knives, pocketknives, hot glue guns, and wire cutters, that can cause harm if handled carelessly. Make safety your number-one priority and give tools your full attention. If you have trouble cutting, your knife or scissors may be dull or the selected material may be too hard; stop immediately and substitute one of the two. *Junior sharpshooters should always be assisted by an adult when handling these tools.*

Always be responsible when constructing and using miniweaponry. It is important that you understand that the author, the publisher, and the bookseller cannot and will not guarantee your safety. When you try the projects described here, you do so at your own risk. They are *not* toys!

SHOOTING RANGE SETUP

Every sharpshooter must have a properly equipped training area to practice his or her MiniWeapon skills. This shooting range can be set up indoors or outdoors, depending on the type of weaponry you are testing. However, for any powerful MiniWeapon capable of firing over long distances, testing should be performed outdoors to reduce property damage and avoid ammunition ricochets.

Follow these simple steps to build your very own custom shooting range. Remember, when practicing on the range, *all shooters and spectators must wear eye protection*.

Step 1

First, locate eye protection for you and any nearby spectators. Then find a large, windowless wall, and move anything of value away from it. Glass, fabrics, and wood can all be damaged by metal dart points, so it's important that these items are removed or covered.

Step 2

Unfold several large cardboard boxes and prop them against the wall to protect it from damage. If the boxes are small and you need to add height, just tape additional bits of cardboard to the existing cardboard barrier.

Step 3

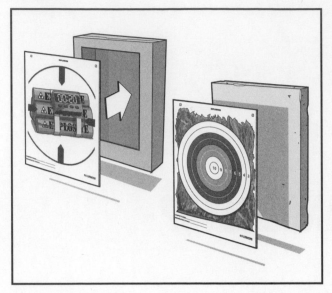

Customize your shooting range for the type of weapons you will be testing. For example, Styrofoam works best for throwing and blow darts, because the dart points can easily stick into the material, making scoring simple. (If you can't locate any Styrofoam, substitute corrugated cardboard.) On the other hand, if you are launching BBs from one of the elastic launchers, you'll need something to catch the projectiles. In that case, cut the front face from a cereal box to create an ammo trap.

Step 4

Tape the ammo trap or Styrofoam onto the cardboard backdrop. The Styrofoam should be at least the size of the paper target. Use thumbtacks to affix a target to the backdrop. Note: In addition to the tear-out targets in this book, additional designs are available in the back of the MiniWeapons building guides or at www.JohnAustinbooks.com.

Step 5

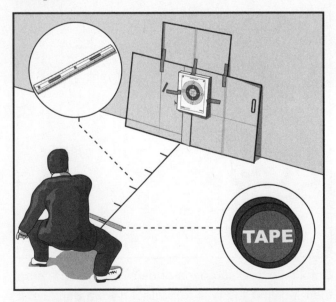

Using a ruler or tape measure, tape off several distances on the floor so that all players can fire from the same mark and the competition is fair. These are called the shooting or throwing lines. To start, we suggest 6-, 10-, 12-, and 15-foot shooting lines.

Step 6

Looks like you are ready test your long- and short-range accuracy. But before each shot, confirm that everyone is behind the firing line.

TARGET SHOOTERS

1

PAPER PICK BLOW GUN

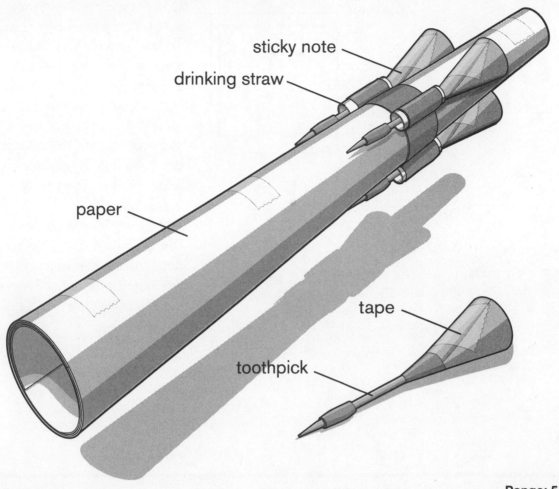

sticky note

drinking straw

paper

tape

toothpick

Range: 5–20 feet

Fabricated from a single sheet of copy paper, modified sticky notes, and wooden toothpicks, the Paper Pick Blowgun is one of the least expensive MiniWeapons of all time. Despite the simple bill of materials, this custom shooter has amazing propulsive power. A burst of air will propel darts at targets up to 20 feet away, making this blowgun an ideal candidate for competitive shooting.

Supplies

1 sheet of copy paper (8.5 inches by 11 inches)
Masking tape or clear tape
1+ square sticky notes (3 inches by 3 inches)
1+ round wooden toothpicks
1 drinking straw

Tools

Safety glasses
Scissors

Step 1

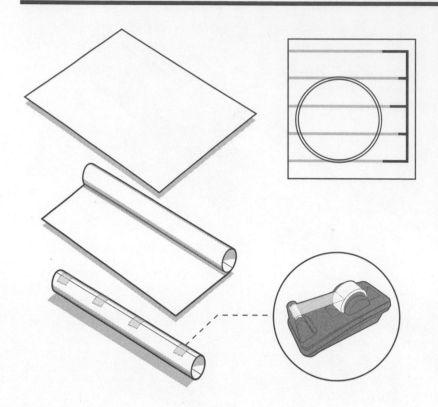

On a flat surface, slowly roll an 8.5-inch-by-11-inch sheet of paper, starting on the 11-inch side, into a tube with a rough diameter of 1 inch. (A large-diameter art marker or wooden dowel can be placed on the paper as a guide to help ensure a uniform width.) Once you have a tight roll, use masking or clear tape to fasten the constructed tube in place. The blowgun assembly is now complete.

Like all MiniWeapons, this design can be modified. A blowgun will produce different results depending on the length of the tube, so experiment by combining sheets of paper or using a larger sheet.

Step 2

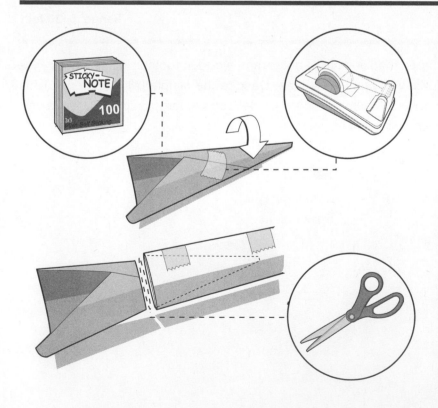

Now it's time to manufacture the fletch of the dart itself. Take one 3-inch-by-3-inch sticky note and roll it into a funnel as shown. Use a small amount of tape to hold the paper in place. Then use additional sticky notes to make three more funnels.

Next, insert one of the constructed funnels into the paper tube, without damaging the tube or funnel. Once the funnel is gently wedged into the tube, use scissors to trim off the extra cone material hanging out of the tube. The blow dart's maximum width is now the same diameter as the blowgun, which will increase the contact surface area when you blow the dart through the tube. Repeat with the remaining three funnels.

Step 3

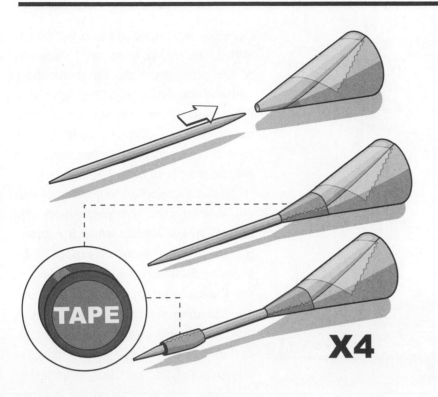

Now slide a single wooden toothpick into the point of one of the paper funnels. If the funnel is too tight at the tip, push the toothpick through from the back. Once it's tucked and centered into the funnel, use tape to secure it as shown. To increase the blow dart's directional accuracy, cut a 3-inch piece of tape in half and tightly wrap it around the toothpick, near the point. You may need to experiment with the amount of tape. Repeat this step until all four darts are completed.

Step 4

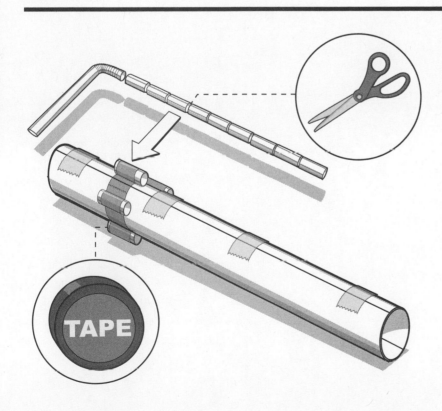

Your Paper Pick Blowgun is now fully functional. However, with this additional step, a tube-mounted dart quiver makes multishot firing easy—and it only takes seconds to construct!

With scissors, cut four small sections from a drinking straw, each roughly 1 inch long. Evenly place each straw segment around the paper tube as shown. Use tape to carefully secure them in place.

Step 5

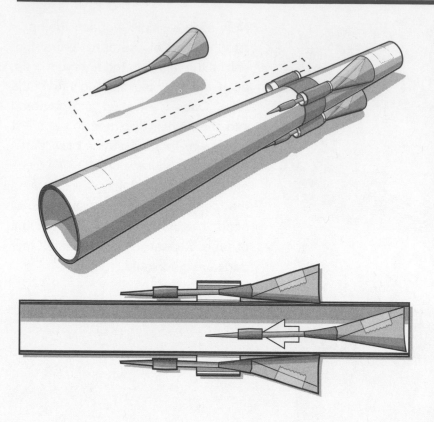

Carefully load the darts into the quiver. When you're ready to fire, stuff one of the projectiles into the paper tube with the toothpick point facing toward the exit, away from the end where your mouth will go. Take a deep breath, lift the blowgun to your lips, aim, and exhale sharply to blow the dart out.

It is very important to remember that you're firing darts from your mouth. *You should never inhale while the blowgun is in your mouth waiting to be launched.* Always be responsible and fire the blowgun in a controlled manner. Wearing safety glasses is a must, as is staying clear of spectators.

Firepower is limited by your respiratory muscles, and accuracy is limited by the blowgun's length and the balance of the darts. Eventually the paper tube will lose its shape and the darts will become dull; replace them as needed.

SEMIAUTOMATIC PEN PISTOL

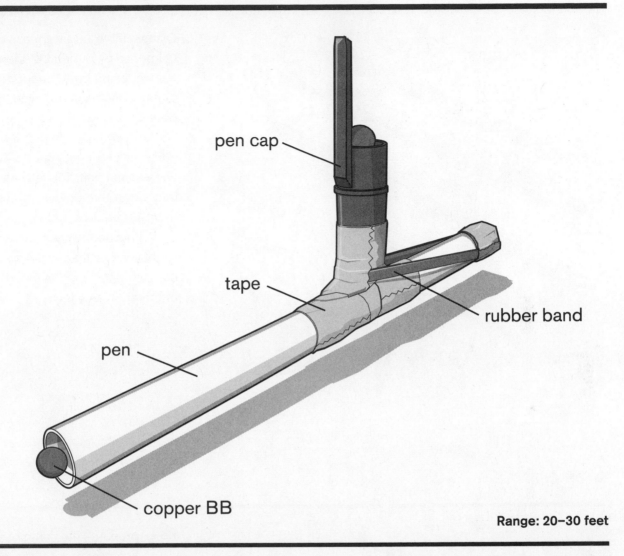

pen cap

tape

pen

rubber band

copper BB

Range: 20–30 feet

Your targets will tremble at the sight of the Semiautomatic Pen Pistol, a pocket-sized machine gun that can unleash fire and brimstone within seconds. If built according to MiniWeapons specifications, the pistol's powerful elastic launcher delivers a single BB each time it's pulled back, automatically chambering a new BB from its pen cap magazine. After a quick assembly, this beauty will instantly become a favorite as it tears targets apart.

Supplies

1 inexpensive mechanical pencil
1 plastic ballpoint pen with cap
Masking tape or duct tape
1 wide rubber band

Tools

Safety glasses
Pliers or thin dowel
Hobby knife
Scissors

Ammo

10+ BBs (pen housing diameter)

Step 1

plunger

pen housing

Disassemble an inexpensive mechanical pencil by pulling out the plunger as shown, using brute strength. Once the plunger has been removed, discard the pencil housing.

Next, remove the tip, ink cartridge, and rear pen-housing cap from a plastic ballpoint pen. You may need a tool—pliers or a thin dowel—to dislodge the rear pen-housing cap.

With a hobby knife, cut a small hole, slightly larger than a BB, in the side of the pen housing, 2½ inches from one of the ends as shown.

Step 2

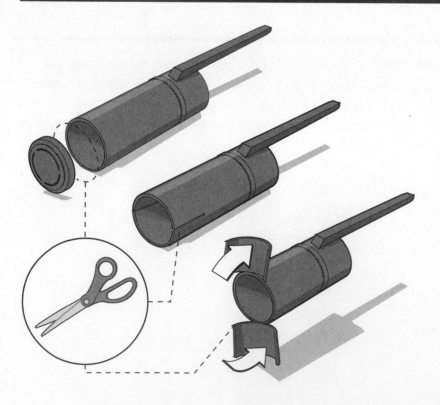

Now prepare the ammo magazine. Using scissors, cut off the pen cap tip as shown.

Next, cut a small slit on either side of the pen cap, with the clip centered between them. Then slowly bend these pen cap segments to a 90-degree angle. You may need to do some additional cutting so the flap details stay in place.

Step 3

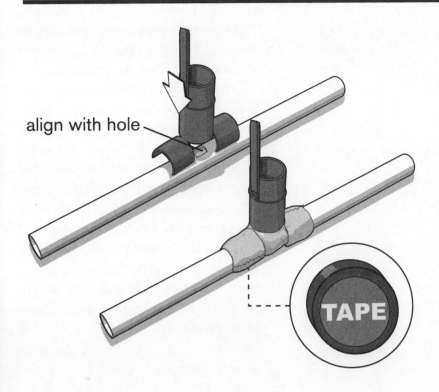

align with hole

TAPE

Now slide the pen cap detail onto the modified pen housing, lining up the pen cap magazine with the BB-sized hole you cut into the pen housing. Once in place, securely tape the cap and pen together.

Test the alignment by placing one small BB into the pen cap. It should roll into the pen housing and then roll out. Adjust the pen housing hole if it's not wide enough.

Step 4

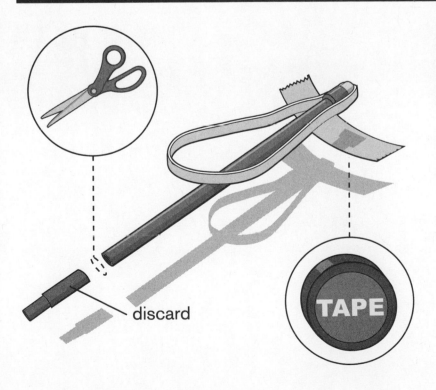

discard

TAPE

Carefully use scissors or a hobby knife to remove the tapered end of the plunger. This will make room for the BB when it sits in the barrel waiting to be launched and will prevent it from falling out.

Next, securely fasten a wide rubber band to the eraser end of the plunger with tape as shown. If the eraser is newer, you may want to cut a slit in the eraser head to slide the rubber band into for additional support.

Step 5

Slide the modified plunger into the pen housing, through the end that is closer to the attached pen cap magazine.

Wrap the attached rubber band around the pen cap housing as shown until you eliminate any slack, then secure the band to the pen assembly using tape. To test the construction, drop a few BBs into the magazine pen cap. Once they're in place, slowly slide the plunger back until one or more of the BBs load into the chamber, then release. Continue that motion until you have unloaded all of your BB ammo. Adjust the construction and rubber band if needed.

Want a larger magazine? Just snap an empty pen housing into the pen cap and load it up.

SPITBALL SHOOTER WITH CLIP

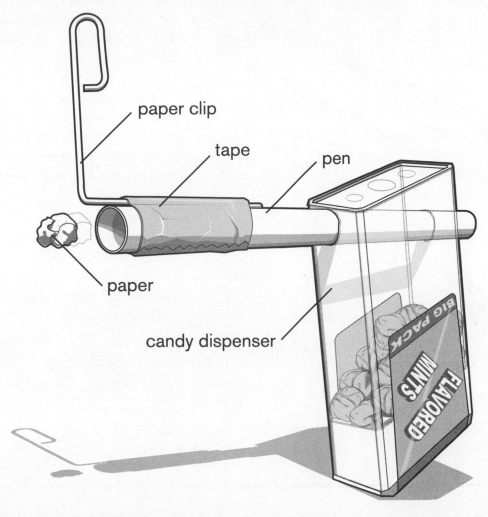

paper clip

tape

pen

paper

candy dispenser

FLAVORED MINTS
BIG PACK

Range: 10–20 feet

Just add saliva! The Spitball Shooter with Clip improves any drooling gunslinger's accuracy and distance by using a fixed-barrel sight and a convenient handle, which also serves as built-in ammo storage. Preload this little devil with a fresh stash of balled-up paper and before you know it you'll be not only plastering paper missiles everywhere but also contributing some much-needed style to spitball warfare.

Supplies

1 plastic ballpoint pen
1 paper clip
1 Tic Tac container
Tape (any kind)

Tools

Safety glasses
Hobby knife (optional)
Pliers

Ammo

1+ square sticky notes (3 inches by 3 inches)

Step 1

discard

First, disassemble an inexpensive pen. Pen designs vary, so make sure the pen housing has a perfectly consistent diameter from front to back, instead of tapering toward the ballpoint tip. (Remove any taper with a hobby knife, if necessary.) Remove the tip, ink cartridge, and rear pen-housing cap. You may need to use pliers to dislodge the housing cap.

Next, to create the ammo, fold one 3-inch-by-3-inch sticky note into nine evenly spaced sections as shown. Using the creased lines as size guides, rip the small sections apart and ball up each of them.

STICKY-NOTE
100
3x3

X9

Step 2

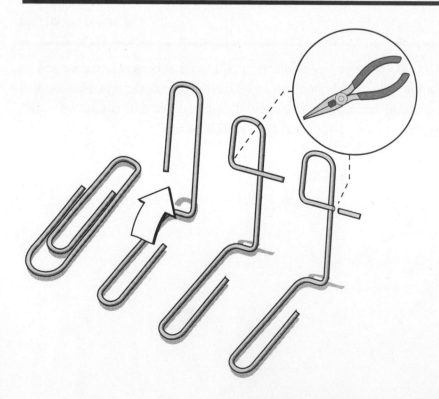

To create a barrel sight, bend one loop of a paper clip to a 90-degree angle as shown. Then use pliers to slowly bend that loop into an eye detail. For aesthetics and safety reasons, use pliers to remove the excess material protruding from the sight detail. The Spitball Shooter sight is complete.

Step 3

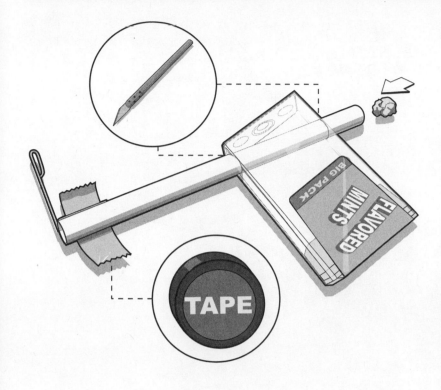

Place the Tic Tac container on a cutting surface. Use the hobby knife to cut two pen-sized holes at a slight angle, near the end opposite the plastic lid, as shown. Slide the pen housing through the holes. Add tape if needed. You have just assembled the Spitball Shooter's handle and barrel.

To complete your MiniWeapon, tape the modified paper clip to the end of the barrel for a sight. Adjust if needed.

To battle, fill the Tic Tac container with wet or dry spitballs and place one of the balled-up pieces of paper into your mouth. Continue to chew and moisten the wad of paper unit it resembles a BB-sized ball. Preload the wet ammo into the rear end of the pen barrel, aim the shooter toward the intended target, and then quickly exhale and watch the wet round spray out.

PEN CAP DART

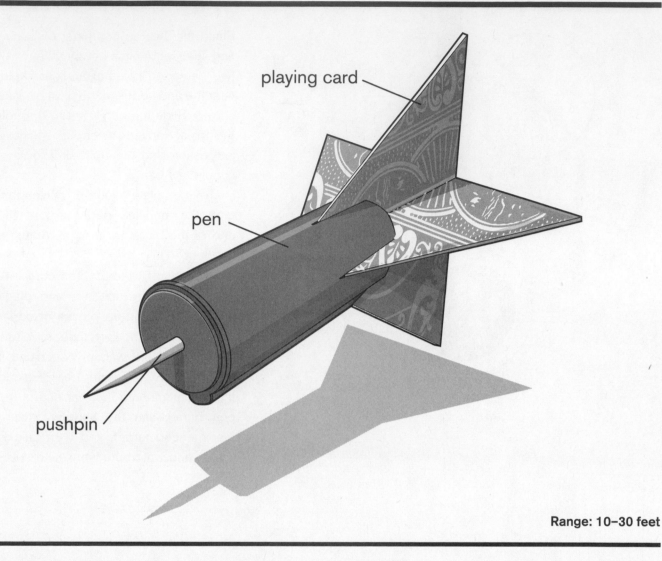

playing card

pen

pushpin

Range: 10–30 feet

The Pen Cap Dart is one of the most durable darts in the MiniWeapons arsenal. It features a heavy-gauge, steel-tip point capable of penetrating almost any target; a high-impact plastic shaft; and durable coated cardstock fins. Sound expensive? It's actually the opposite, created from discarded office items. It's also quick to assemble; you should have no problem creating a custom set of darts before tournament time.

Supplies

1 plastic pen cap
1 pushpin
1 playing card

Tools

Safety glasses
Scissors
Hot glue gun (optional)

Step 1

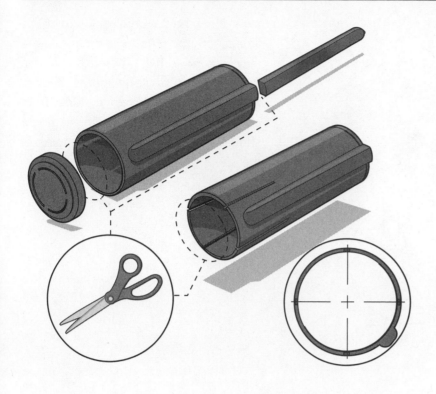

With scissors, remove both the clip and the enclosed tip from a pen cap, as shown. Then cut four evenly spaced half-inch slits onto the walls of the pen cap. The Pen Cap Dart plastic shaft is complete.

Step 2

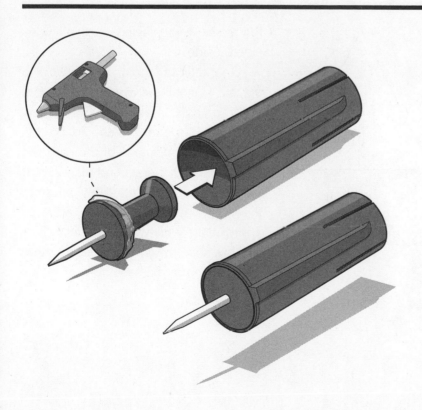

Now dab a small amount of hot glue inside the pen cap and then slide the pushpin (point out) into the pen cap opening, opposite the newly cut slits. Depending on the diameter of both the pen cap and the pushpin, hot glue may not be needed; you may be able to simply wrap clear tape around the pushpin handle to create a snug fit.

Step 3

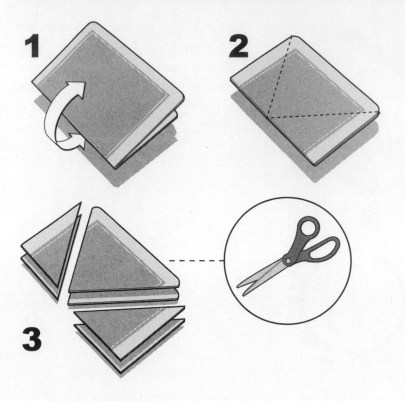

Now it's time to construct the dart's fins, also known as flights. First, fold a playing card in half. Then use scissors to cut out a triangle shape, as shown. Remove the extra card material from both sides. When you're finished, you should have two separate triangles of the same size.

Step 4

Place the two playing card triangles side by side.

On the first triangle, use scissors to cut out a small slit from the top point of the triangle to about halfway *down*. The width of the slit should be the same as the thickness of the card but not bigger. On the second triangle, cut out a small slit of the same width from the midpoint of the bottom edge to approximately halfway *up* the triangle.

Now slide the two triangles together to form the rear fin assembly.

Step 5

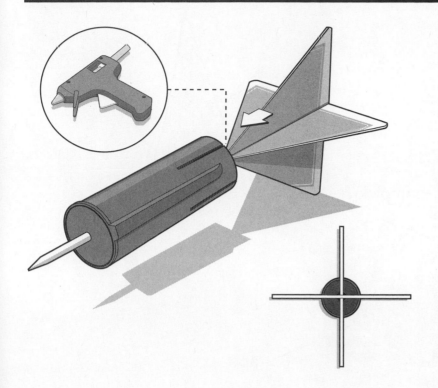

Now it's time to combine both dart assemblies. Dab a small amount of hot glue (optional) onto the tip of the fin assembly as indicated in the illustration. Then slide that fin assembly into the pen cap, aligning the fins with the four slots you cut earlier.

It is important to remember that a dart is equipped with a dangerous point at the end and *is not meant for living targets*. Malfunctions do occur, so use the utmost caution when tossing homemade darts. See the Shooting Range Setup (page vii) for a controlled environment in which to practice.

TOOTHPICK TAPE DART

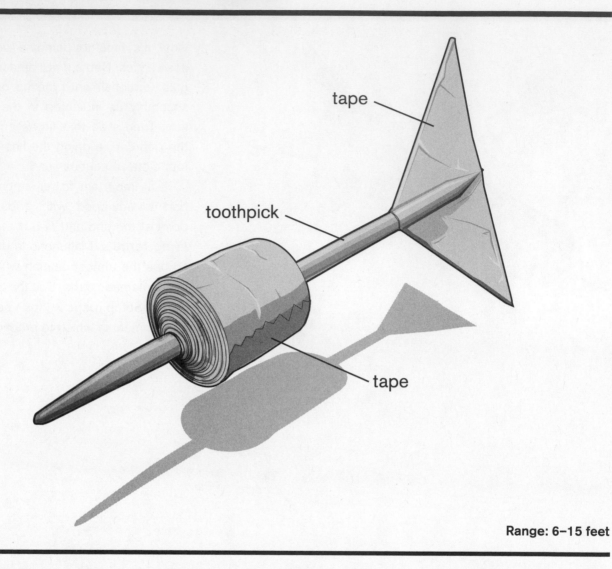

tape

toothpick

tape

Range: 6–15 feet

The Toothpick Tape Dart is the ultimate answer for those in search of a very quick MiniWeapon build. Its straightforward construction and short-distance accuracy make it the perfect option for mass production. Plus, the Toothpick Tape Dart encourages customization by adjusting or adding fins and increasing or decreasing the shaft weight. With a few tweaks, you'll be skewing bull's-eye after bull's-eye.

Supplies

Masking tape or duct tape
1 round wooden toothpick

Tools

Safety glasses
Scissors

Step 1

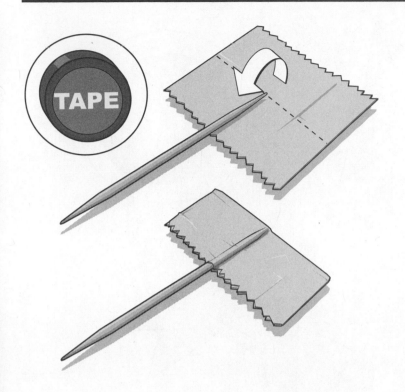

Start by creating fins for the dart. Depending on the width of your masking or duct tape, you probably will be folding it the long way or in half (as shown). Tear off a medium-sized piece of tape (2 inches long or longer) and place it on a flat surface with the sticky side up. Then center the round toothpick ¾ inch onto the sticky side as shown. Once in place, fold the tape so that the finished fold is approximately 1 inch long by 1½ inches wide, with the toothpick centered.

Step 2

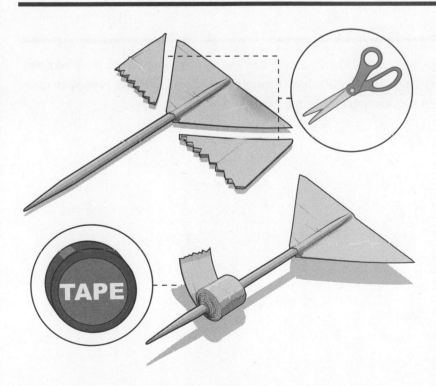

Use scissors to make two diagonal cuts in the attached tape as shown, creating a single triangle. The triangle point should point toward the toothpick point. The fin assembly is complete.

You will need to add weight to the dart point to increase its accuracy and distance. To do this, take a 12-inch section of tape and trim its width to ½ inch using scissors (or carefully tear it). Once you have a thin strip, tightly wind it around the toothpick about ½ inch from the point. You may need to decrease or increase the amount of tape depending on manufacture, so be sure to experiment.

Remember that *even homemade darts are dangerous and are not meant for living targets*. Always use common sense when throwing darts.

Alternate Construction

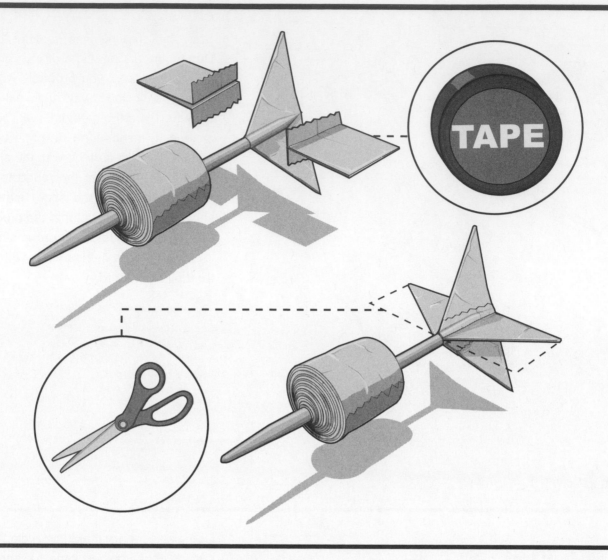

If you find the dart design a little unstable, you can add two additional fins to the side. To do this, fold two small pieces of tape with each sticky end flipped up at a 90-degree angle. Evenly space both pieces of tape onto the sides of the original fins, then use scissors to create the triangle detail.

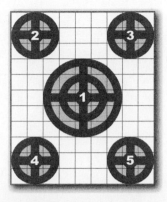

BASIC TARGETS

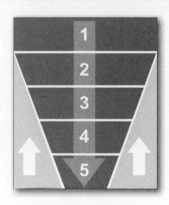

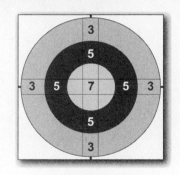

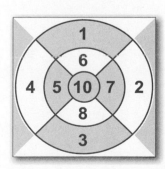

MiniWeapons of Mass Destruction Targets, © Austin Design, Inc.

Competitor(s) _____   Date _____

MiniWeapon Used _____

BASIC BULL'S-EYE RULES

Basic Rules ★

Before the game starts, the players decide on the distance of the throwing line, between 5 and 12 feet from the target. Alternating turns, each player fires three times at the target from the throwing line. When each player is finished shooting, add his or her target scores and continue the game until everyone has had three turns. The shooter with the highest total score wins.

Marksman Rules ★★★

Because you are professional sharpshooters, fire your MiniWeapons from at least 10 feet away. Remember to alternate turns. The first marksman to score a cumulative 50 or more points wins this game!

Competitor(s) _____   Date _____

MiniWeapon Used _____

BASIC BULL'S-EYE RULES

Basic Rules ★

Before the game starts, the players decide on the distance of the throwing line, between 5 and 12 feet from the target. Alternating turns, each player fires three times at the target from the throwing line. When each player is finished shooting, add his or her target scores and continue the game until everyone has had three turns. The shooter with the highest total score wins.

Marksman Rules ★ ★ ★

Because you are professional sharpshooters, fire your MiniWeapons from at least 10 feet away. Remember to alternate turns. The first marksman to score a cumulative 50 or more points wins this game!

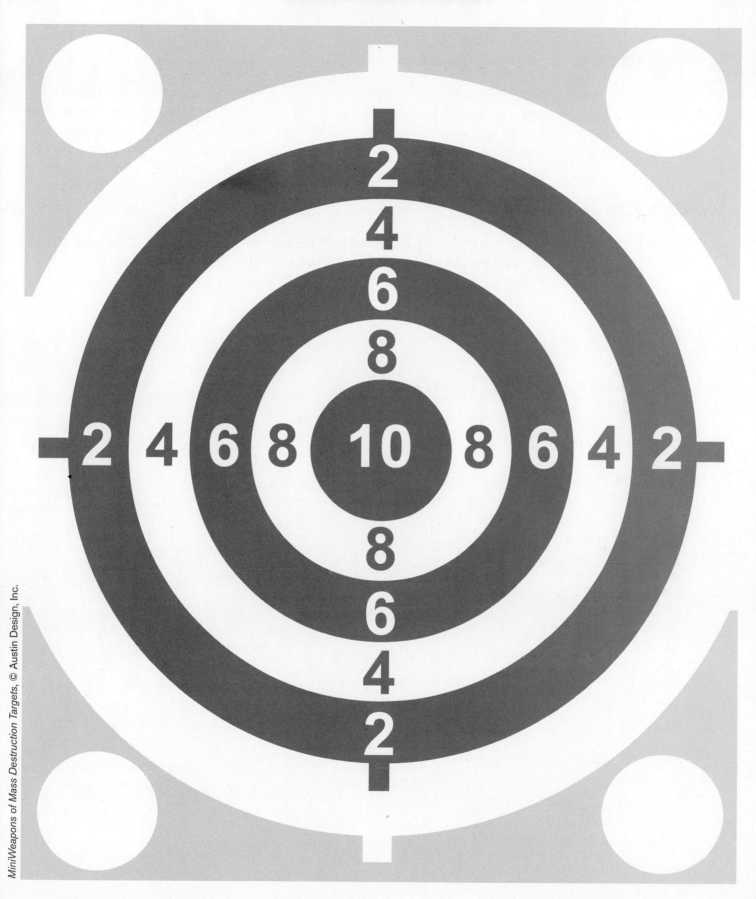

Competitor(s) _____ Date _____

MiniWeapon Used _____

BASIC BULL'S-EYE RULES

Basic Rules ★

Before the game starts, the players decide on the distance of the throwing line, between 5 and 12 feet from the target. Alternating turns, each player fires three times at the target from the throwing line. When each player is finished shooting, add his or her target scores and continue the game until everyone has had three turns. The shooter with the highest total score wins.

Marksman Rules ★★★

Because you are professional sharpshooters, fire your MiniWeapons from at least 10 feet away. Remember to alternate turns. The first marksman to score a cumulative 50 or more points wins this game!

Competitor(s) _____   Date _____

MiniWeapon Used _____

BASIC BULL'S-EYE RULES

Basic Rules ★

Before the game starts, the players decide on the distance of the throwing line, between 5 and 12 feet from the target. Alternating turns, each player fires three times at the target from the throwing line. When each player is finished shooting, add his or her target scores and continue the game until everyone has had three turns. The shooter with the highest total score wins.

Marksman Rules ★★★

Because you are professional sharpshooters, fire your MiniWeapons from at least 10 feet away. Remember to alternate turns. The first marksman to score a cumulative 50 or more points wins this game!

Competitor(s) _____ Date _____

MiniWeapon Used _____

BASIC BULL'S-EYE RULES

Basic Rules ★

Before the game starts, the players decide on the distance of the throwing line, between 5 and 12 feet from the target. Alternating turns, each player fires three times at the target from the throwing line. When each player is finished shooting, add his or her target scores and continue the game until everyone has had three turns. The shooter with the highest total score wins.

Marksman Rules ★★★

Because you are professional sharpshooters, fire your MiniWeapons from at least 10 feet away. Remember to alternate turns. The first marksman to score a cumulative 50 or more points wins this game!

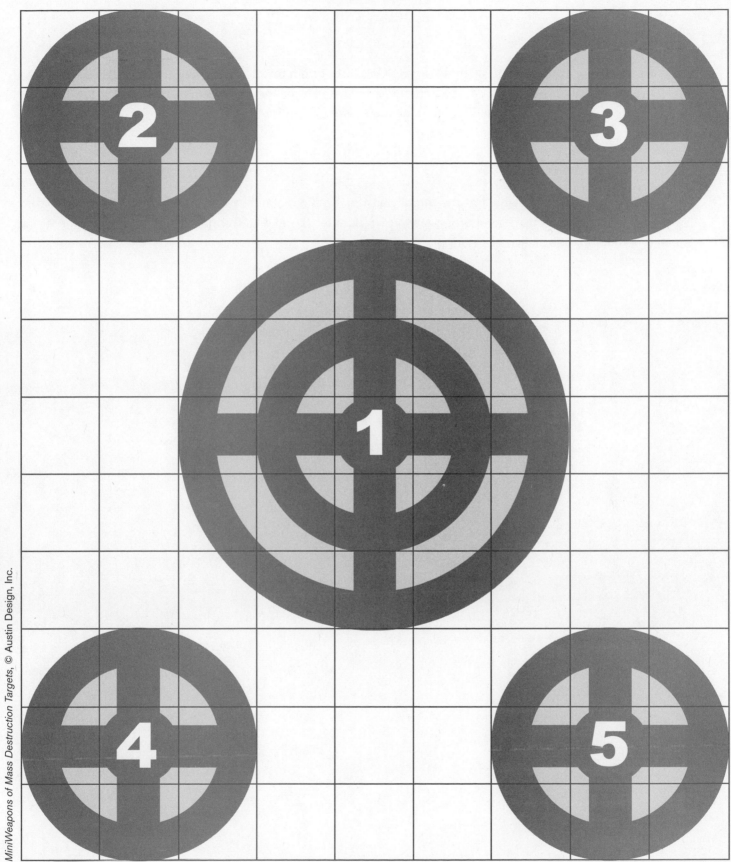

Competitor(s) _____   Date _____

MiniWeapon Used _____

FIVE SHOT RULES

Basic Rules ★

Alternating turns, each player fires three times at the target from 5 to 12 feet away. Each circle has a labeled point value; add up those points at the end of each player's turn. The game ends when each player has had three turns. The shooter with the highest total score wins.

Marksman Rules ★★★

Each marksman takes five shots at the target per turn, from a distance predetermined by the players. They must hit each target in numerical order, staring with center target #1. The game continues until one of the marksmen completes all five hits during a single turn.

MiniWeapons of Mass Destruction Targets, © Austin Design, Inc.

Competitor(s) _____   Date _____

MiniWeapon Used _____

FIVE SHOT RULES

Basic Rules ★

Alternating turns, each player fires three times at the target from 5 to 12 feet away. Each circle has a labeled point value; add up those points at the end of each player's turn. The game ends when each player has had three turns. The shooter with the highest total score wins.

Marksman Rules ★ ★ ★

Each marksman takes five shots at the target per turn, from a distance predetermined by the players. They must hit each target in numerical order, staring with center target #1. The game continues until one of the marksmen completes all five hits during a single turn.

MiniWeapons of Mass Destruction Targets, © Austin Design, Inc.

Competitor(s) _____   Date _____

MiniWeapon Used _____

FIVE SHOT RULES

Basic Rules ★

Alternating turns, each player fires three times at the target from 5 to 12 feet away. Each circle has a labeled point value; add up those points at the end of each player's turn. The game ends when each player has had three turns. The shooter with the highest total score wins.

Marksman Rules ★★★

Each marksman takes five shots at the target per turn, from a distance predetermined by the players. They must hit each target in numerical order, staring with center target #1. The game continues until one of the marksmen completes all five hits during a single turn.

MiniWeapons of Mass Destruction Targets, © Austin Design, Inc.

Competitor(s) _____   Date _____

MiniWeapon Used _____

FIVE SHOT RULES

Basic Rules ★

Alternating turns, each player fires three times at the target from 5 to 12 feet away. Each circle has a labeled point value; add up those points at the end of each player's turn. The game ends when each player has had three turns. The shooter with the highest total score wins.

Marksman Rules ★★★

Each marksman takes five shots at the target per turn, from a distance predetermined by the players. They must hit each target in numerical order, staring with center target #1. The game continues until one of the marksmen completes all five hits during a single turn.

FIVE SHOT

Competitor(s) _____   Date _____

FIVE SHOT RULES

Basic Rules ★

Alternating turns, each player fires three times at the target from 5 to 12 feet away. Each circle has a labeled point value; add up those points at the end of each player's turn. The game ends when each player has had three turns. The shooter with the highest total score wins.

Marksman Rules ★ ★ ★

Each marksman takes five shots at the target per turn, from a distance predetermined by the players. They must hit each target in numerical order, staring with center target #1. The game continues until one of the marksmen completes all five hits during a single turn.

Competitor(s) _____   Date _____

MiniWeapon Used _____

DOWN THE ROW RULES

Basic Rules ★

Alternating turns, each player fires three times at the target from 5 to 12 feet away. Each bar has a labeled point value; add up those points at the end of each player's turn. The game ends when each player has had three turns. The shooter with the highest total score wins.

Marksman Rules ★★★

This will take a marksman! After players decide on the firing distance, the first shooter aims for the top of the target and then continues in numerical order down the ladder until he or she reaches #5. If the shooter misses the target or hits an "up arrow" zone, it's the next player's turn. The first player to complete this task during a single sequence wins.

1

2

3

4

5

Competitor(s) _____ Date _____

MiniWeapon Used _____

DOWN THE ROW RULES

Basic Rules ★

Alternating turns, each player fires three times at the target from 5 to 12 feet away. Each bar has a labeled point value; add up those points at the end of each player's turn. The game ends when each player has had three turns. The shooter with the highest total score wins.

Marksman Rules ★ ★ ★

This will take a marksman! After players decide on the firing distance, the first shooter aims for the top of the target and then continues in numerical order down the ladder until he or she reaches #5. If the shooter misses the target or hits an "up arrow" zone, it's the next player's turn. The first player to complete this task during a single sequence wins.

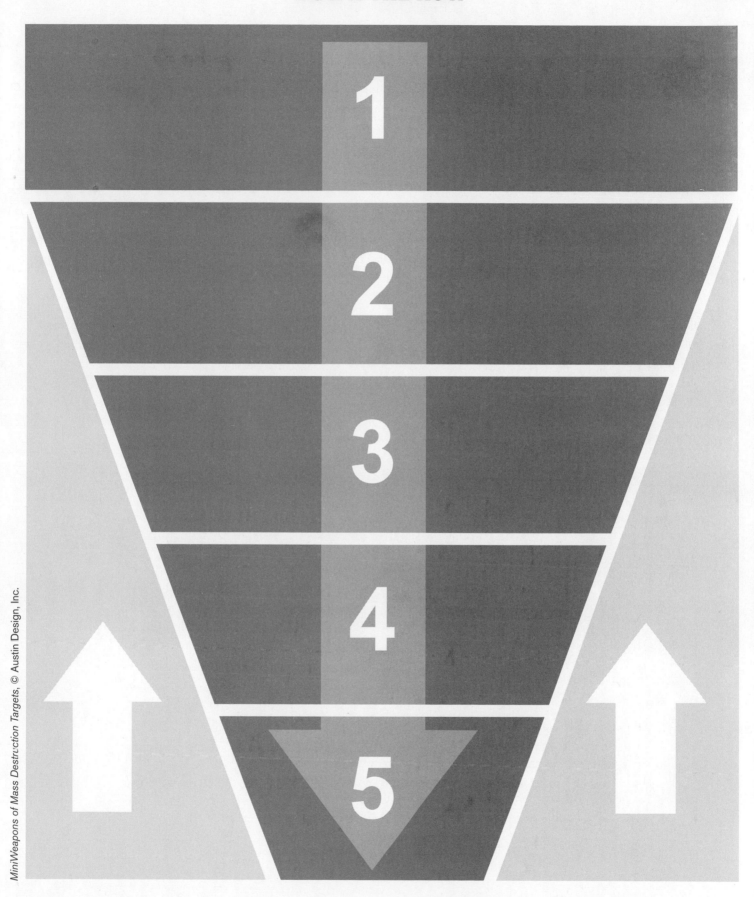

Competitor(s) _____ Date _____

MiniWeapon Used _____

DOWN THE ROW RULES

Basic Rules ★

Alternating turns, each player fires three times at the target from 5 to 12 feet away. Each bar has a labeled point value; add up those points at the end of each player's turn. The game ends when each player has had three turns. The shooter with the highest total score wins.

Marksman Rules ★ ★ ★

This will take a marksman! After players decide on the firing distance, the first shooter aims for the top of the target and then continues in numerical order down the ladder until he or she reaches #5. If the shooter misses the target or hits an "up arrow" zone, it's the next player's turn. The first player to complete this task during a single sequence wins.

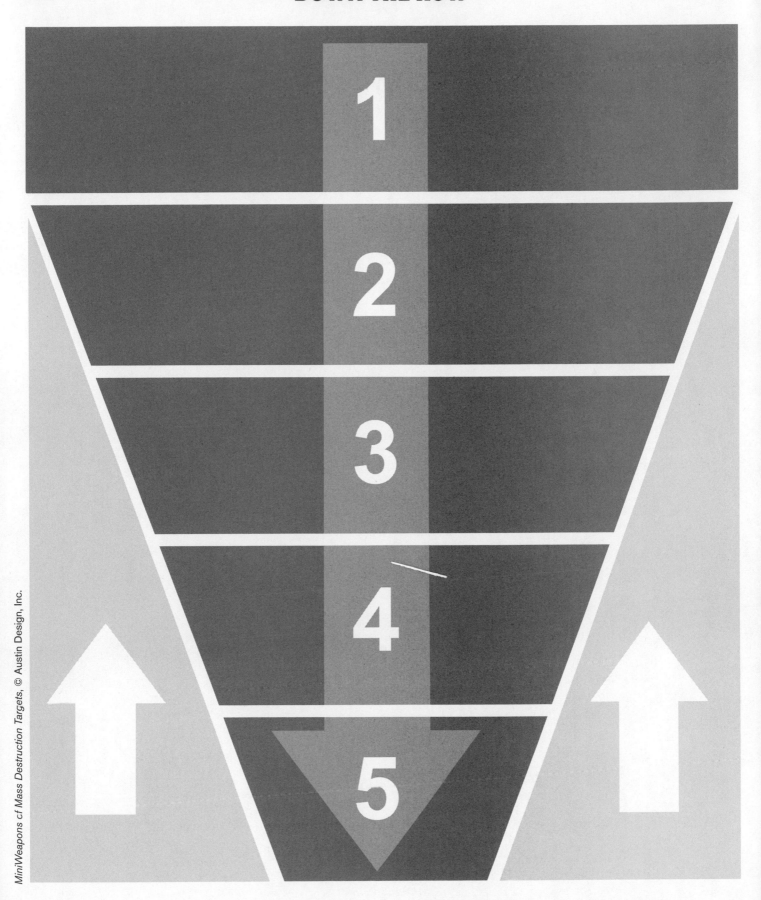

Competitor(s) _____ Date _____

MiniWeapon Used _____

DOWN THE ROW RULES

Basic Rules ★

Alternating turns, each player fires three times at the target from 5 to 12 feet away. Each bar has a labeled point value; add up those points at the end of each player's turn. The game ends when each player has had three turns. The shooter with the highest total score wins.

Marksman Rules ★ ★ ★

This will take a marksman! After players decide on the firing distance, the first shooter aims for the top of the target and then continues in numerical order down the ladder until he or she reaches #5. If the shooter misses the target or hits an "up arrow" zone, it's the next player's turn. The first player to complete this task during a single sequence wins.

DOWN THE ROW

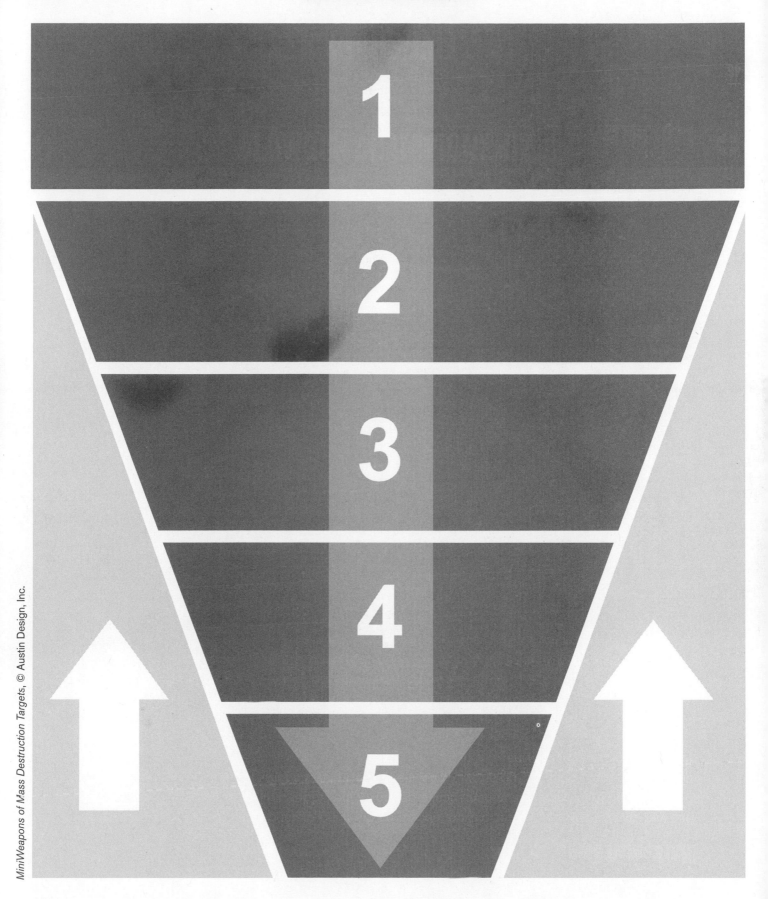

Competitor(s) _____ Date _____

MiniWeapon Used _____

DOWN THE ROW RULES

Basic Rules ★

Alternating turns, each player fires three times at the target from 5 to 12 feet away. Each bar has a labeled point value; add up those points at the end of each player's turn. The game ends when each player has had three turns. The shooter with the highest total score wins.

Marksman Rules ★★★

This will take a marksman! After players decide on the firing distance, the first shooter aims for the top of the target and then continues in numerical order down the ladder until he or she reaches #5. If the shooter misses the target or hits an "up arrow" zone, it's the next player's turn. The first player to complete this task during a single sequence wins.

BLOWGUN CHAMPIONSHIP

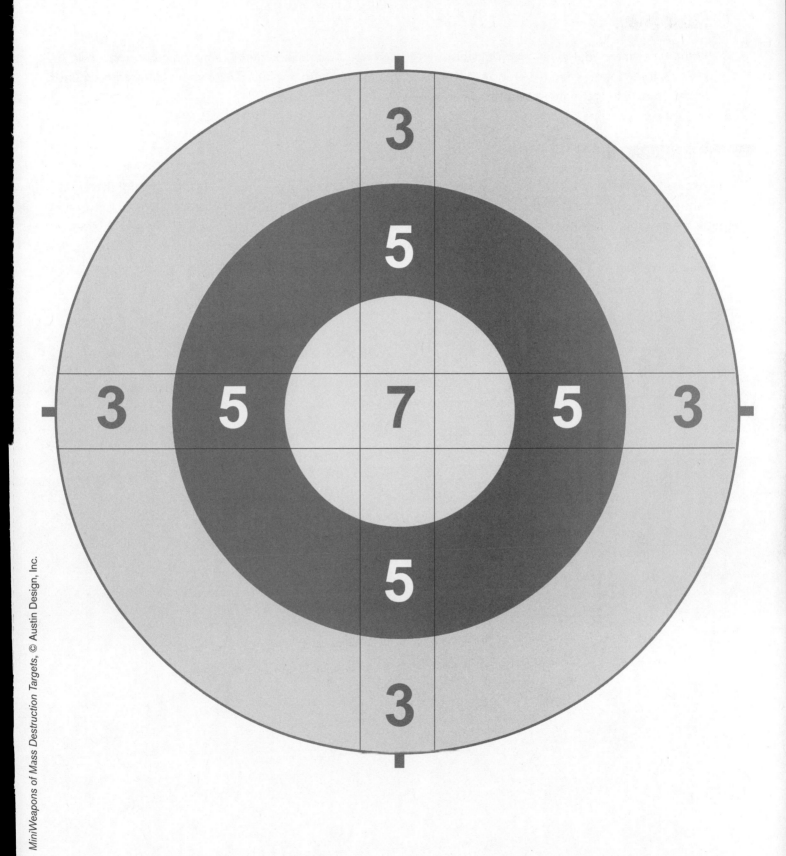

Competitor(s) _____ Date _____

MiniWeapon Used _____

BLOWGUN CHAMPIONSHIP RULES

Basic Rules ★

From 10 or more feet away, each player takes three shots at the Blowgun Championship target. Add the player's total points when retrieving the darts and then continue to the next player's turn. After three rounds, the shooter with the highest total score wins.

Marksman Rules ★ ★ ★

Similar to the basketball game "PIG," this game is called "TARGET." You may fire at the target from any distance and location, and if you hit the target anywhere, the next player must attempt the same shot from the same position. If the other player does not successfully hit the target, he or she earns the letter *T* in the word *TARGET*. Continue until one of the players is eliminated by spelling the whole word. If player 1 misses the target, then player 2 is the leader.

BLOWGUN CHAMPIONSHIP

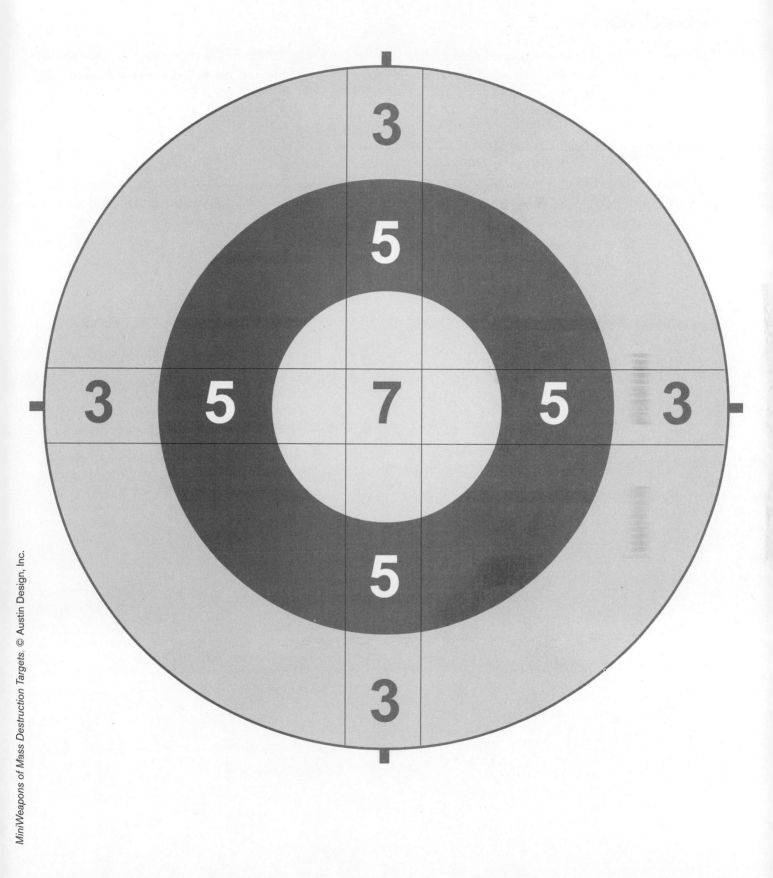

Competitor(s) _____   Date _____

MiniWeapon Used _____

BLOWGUN CHAMPIONSHIP RULES

Basic Rules ★

From 10 or more feet away, each player takes three shots at the Blowgun Championship target. Add the player's total points when retrieving the darts and then continue to the next player's turn. After three rounds, the shooter with the highest total score wins.

Marksman Rules ★★★

Similar to the basketball game "PIG," this game is called "TARGET." You may fire at the target from any distance and location, and if you hit the target anywhere, the next player must attempt the same shot from the same position. If the other player does not successfully hit the target, he or she earns the letter *T* in the word *TARGET*. Continue until one of the players is eliminated by spelling the whole word. If player 1 misses the target, then player 2 is the leader.

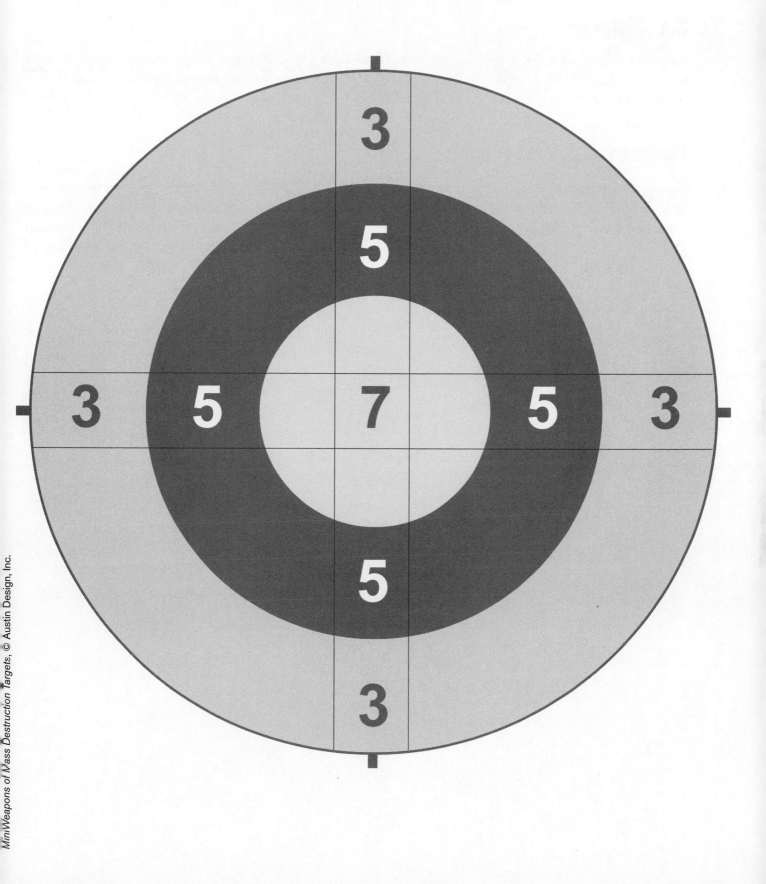

Competitor(s) _____   Date _____

MiniWeapon Used _____

BLOWGUN CHAMPIONSHIP RULES

Basic Rules ★

From 10 or more feet away, each player takes three shots at the Blowgun Championship target. Add the player's total points when retrieving the darts and then continue to the next player's turn. After three rounds, the shooter with the highest total score wins.

Marksman Rules ★★★

Similar to the basketball game "PIG," this game is called "TARGET." You may fire at the target from any distance and location, and if you hit the target anywhere, the next player must attempt the same shot from the same position. If the other player does not successfully hit the target, he or she earns the letter *T* in the word *TARGET*. Continue until one of the players is eliminated by spelling the whole word. If player 1 misses the target, then player 2 is the leader.

BLOWGUN CHAMPIONSHIP

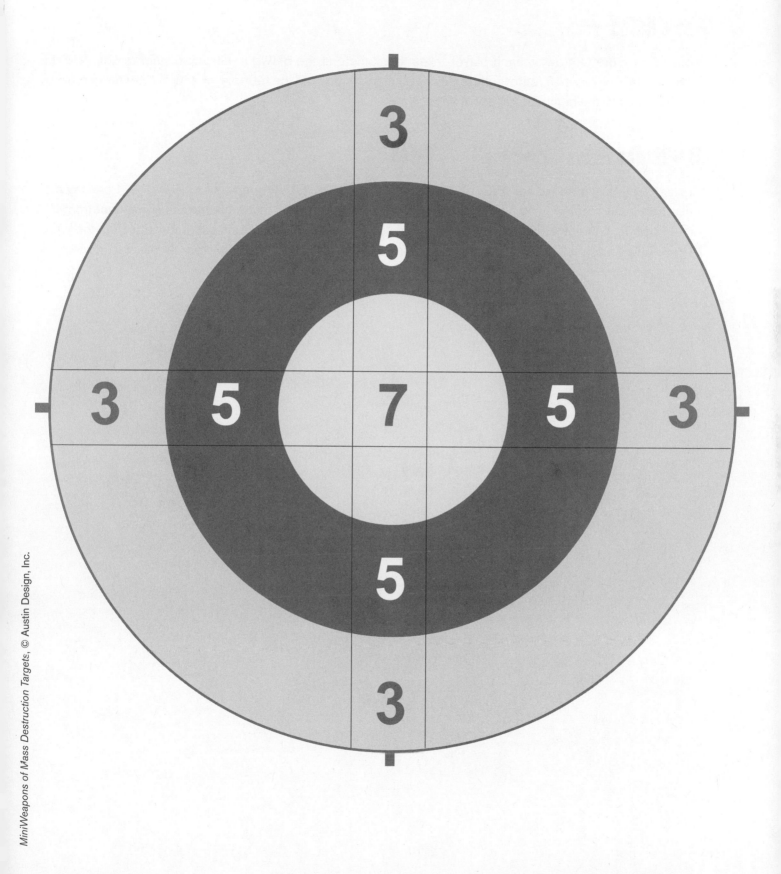

Competitor(s) _____   Date _____

MiniWeapon Used _____

BLOWGUN CHAMPIONSHIP RULES

Basic Rules ★

From 10 or more feet away, each player takes three shots at the Blowgun Championship target. Add the player's total points when retrieving the darts and then continue to the next player's turn. After three rounds, the shooter with the highest total score wins.

Marksman Rules ★ ★ ★

Similar to the basketball game "PIG," this game is called "TARGET." You may fire at the target from any distance and location, and if you hit the target anywhere, the next player must attempt the same shot from the same position. If the other player does not successfully hit the target, he or she earns the letter *T* in the word *TARGET*. Continue until one of the players is eliminated by spelling the whole word. If player 1 misses the target, then player 2 is the leader.

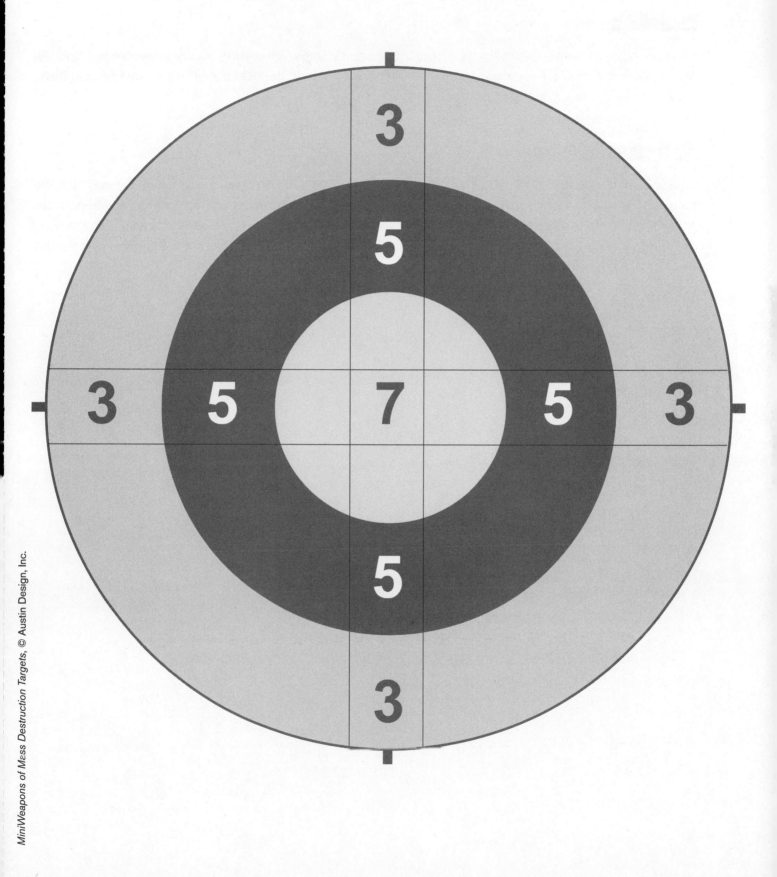

Competitor(s) _____ Date _____

MiniWeapon Used _____

BLOWGUN CHAMPIONSHIP RULES

Basic Rules ★

From 10 or more feet away, each player takes three shots at the Blowgun Championship target. Add the player's total points when retrieving the darts and then continue to the next player's turn. After three rounds, the shooter with the highest total score wins.

Marksman Rules ★ ★ ★

Similar to the basketball game "PIG," this game is called "TARGET." You may fire at the target from any distance and location, and if you hit the target anywhere, the next player must attempt the same shot from the same position. If the other player does not successfully hit the target, he or she earns the letter *T* in the word *TARGET*. Continue until one of the players is eliminated by spelling the whole word. If player 1 misses the target, then player 2 is the leader.

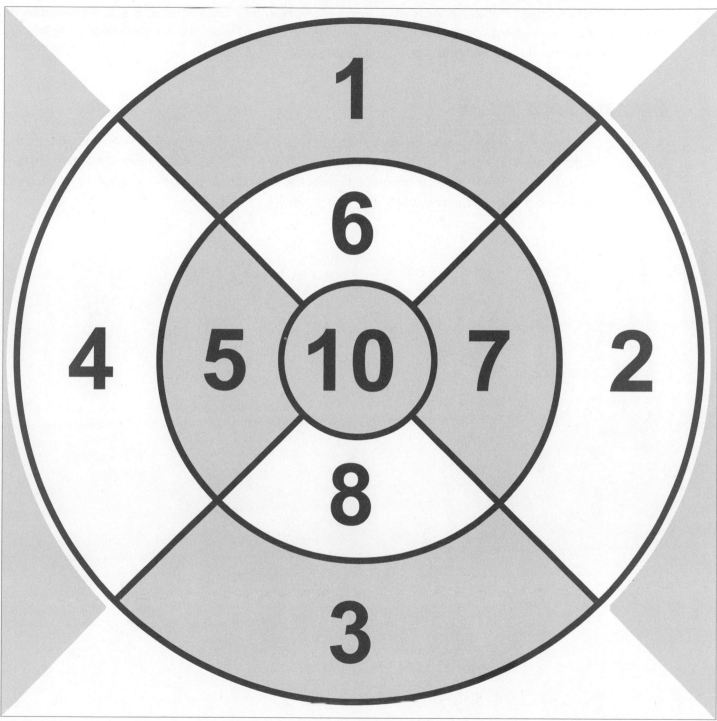

Competitor(s) _____

Date _____

MiniWeapon Used _____

AROUND THE WORLD RULES

Basic Rules ★

Use this target as a basic bull's-eye. Alternating turns, each player fires three times at the target from 5 to 12 feet away. When each shooter is finished, add up his or her score and continue the game until everyone has had three turns. The shooter with the highest total score wins.

Marksman Rules ★ ★ ★

After everyone agrees on a designated distance from the target, each player tries to hit each number on the target in numeric order, starting at 1 and ending with the bull's-eye, 10. If a player misses a number, the turn passes to the next player. The first player to complete this task wins. Game can be won in a single sequence (difficult) or by keeping track of the previously made shots (easy).

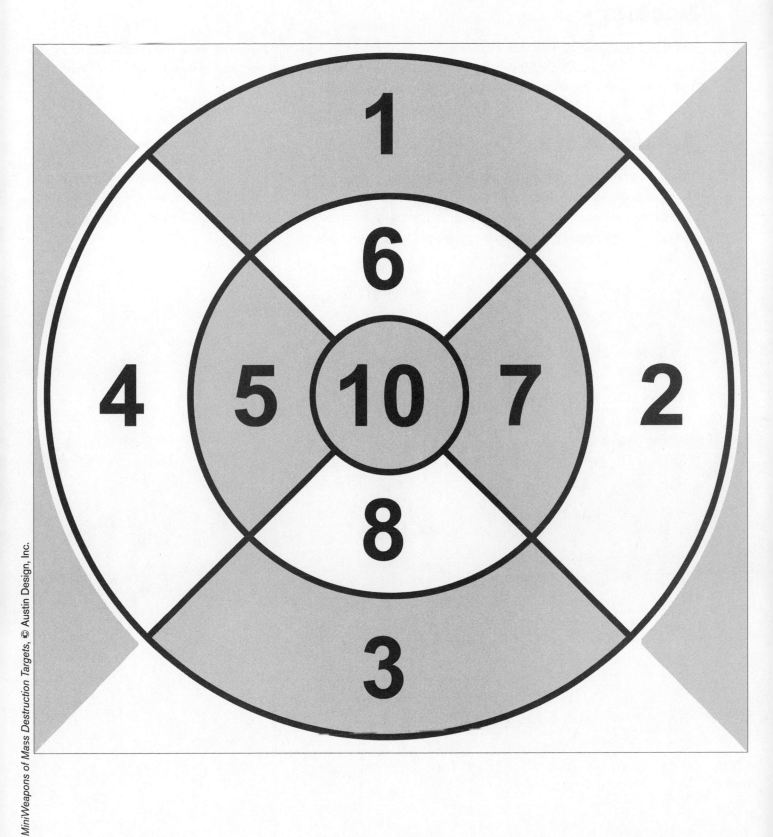

Competitor(s) _____ Date _____

MiniWeapon Used _____

AROUND THE WORLD RULES

Basic Rules ★

Use this target as a basic bull's-eye. Alternating turns, each player fires three times at the target from 5 to 12 feet away. When each shooter is finished, add up his or her score and continue the game until everyone has had three turns. The shooter with the highest total score wins.

Marksman Rules ★★★

After everyone agrees on a designated distance from the target, each player tries to hit each number on the target in numeric order, starting at 1 and ending with the bull's-eye, 10. If a player misses a number, the turn passes to the next player. The first player to complete this task wins. Game can be won in a single sequence (difficult) or by keeping track of the previously made shots (easy).

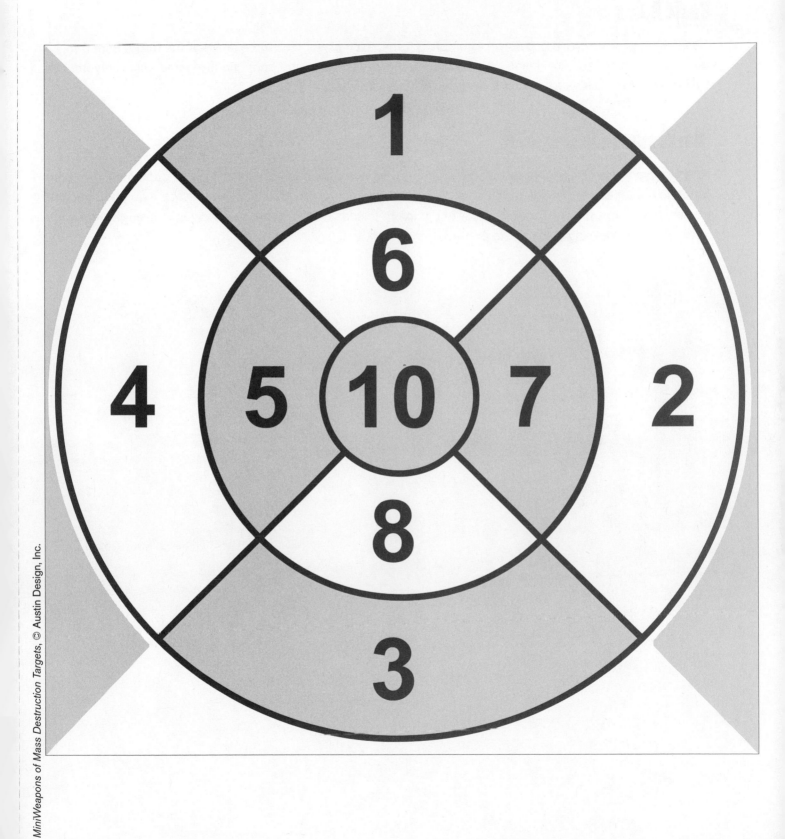

MiniWeapons of Mass Destruction Targets, © Austin Design, Inc.

Competitor(s) _____   Date _____

MiniWeapon Used _____

AROUND THE WORLD RULES

Basic Rules ★

Use this target as a basic bull's-eye. Alternating turns, each player fires three times at the target from 5 to 12 feet away. When each shooter is finished, add up his or her score and continue the game until everyone has had three turns. The shooter with the highest total score wins.

Marksman Rules ★★★

After everyone agrees on a designated distance from the target, each player tries to hit each number on the target in numeric order, starting at 1 and ending with the bull's-eye, 10. If a player misses a number, the turn passes to the next player. The first player to complete this task wins. Game can be won in a single sequence (difficult) or by keeping track of the previously made shots (easy).

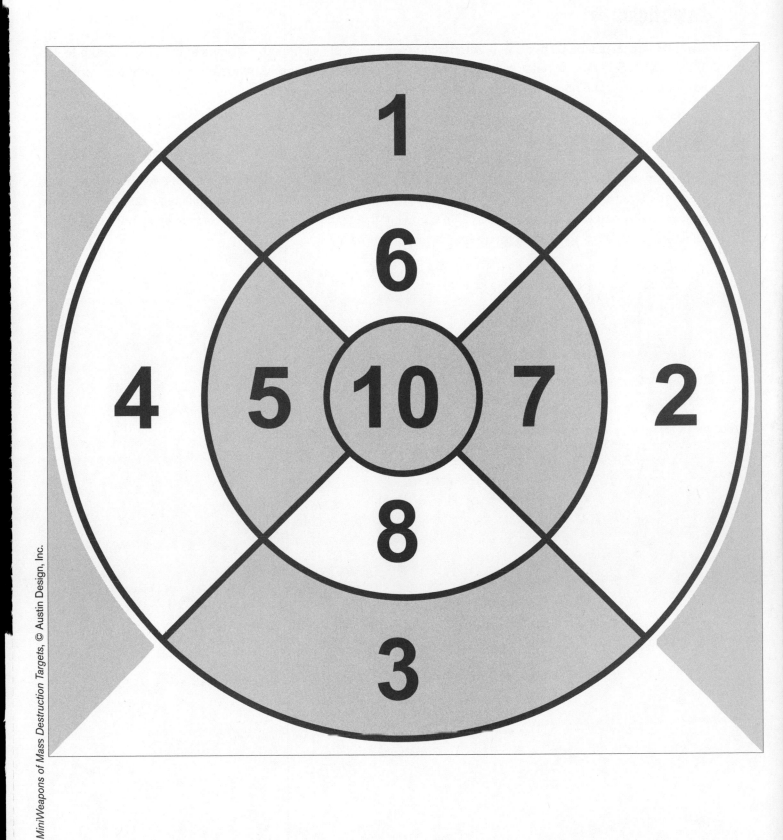

Competitor(s) _____ Date _____

MiniWeapon Used _____

AROUND THE WORLD RULES

Basic Rules ★

Use this target as a basic bull's-eye. Alternating turns, each player fires three times at the target from 5 to 12 feet away. When each shooter is finished, add up his or her score and continue the game until everyone has had three turns. The shooter with the highest total score wins.

Marksman Rules ★★★

After everyone agrees on a designated distance from the target, each player tries to hit each number on the target in numeric order, starting at 1 and ending with the bull's-eye, 10. If a player misses a number, the turn passes to the next player. The first player to complete this task wins. Game can be won in a single sequence (difficult) or by keeping track of the previously made shots (easy).

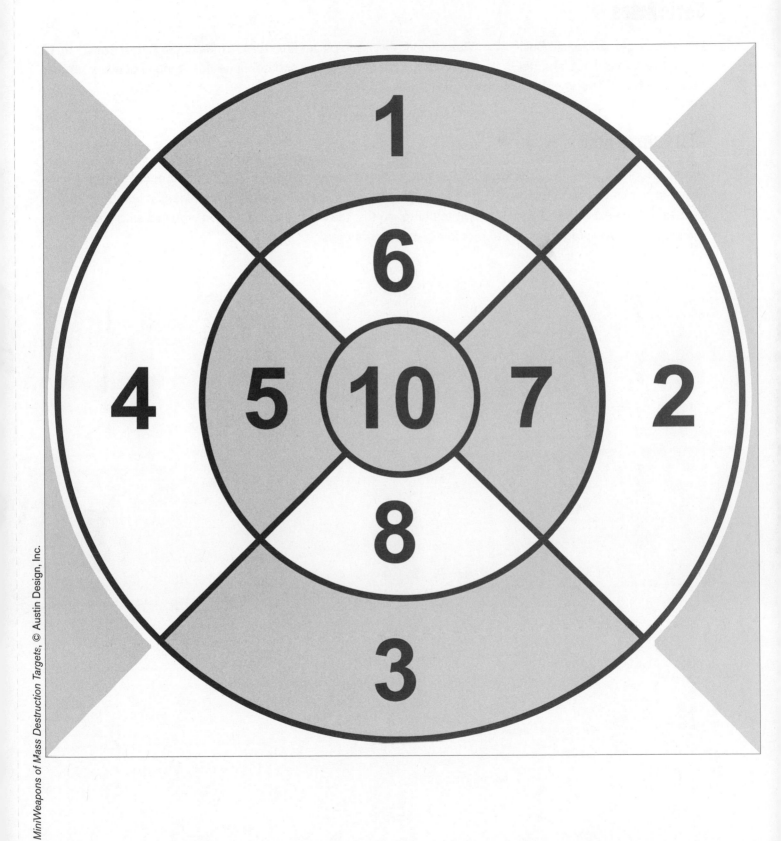

MiniWeapons of Mass Destruction Targets, © Austin Design, Inc.

Competitor(s) _____   Date _____

MiniWeapon Used _____

AROUND THE WORLD RULES

Basic Rules ★

Use this target as a basic bull's-eye. Alternating turns, each player fires three times at the target from 5 to 12 feet away. When each shooter is finished, add up his or her score and continue the game until everyone has had three turns. The shooter with the highest total score wins.

Marksman Rules ★ ★ ★

After everyone agrees on a designated distance from the target, each player tries to hit each number on the target in numeric order, starting at 1 and ending with the bull's-eye, 10. If a player misses a number, the turn passes to the next player. The first player to complete this task wins. Game can be won in a single sequence (difficult) or by keeping track of the previously made shots (easy).

	20			19			18			17			16			15			Bull		
P1	○	○	○	○	○	○	○	○	○	○	○	○	○	○	○	○	○	○	○	○	○
P2	○	○	○	○	○	○	○	○	○	○	○	○	○	○	○	○	○	○	○	○	○
P3	○	○	○	○	○	○	○	○	○	○	○	○	○	○	○	○	○	○	○	○	○

Competitor(s) _____   Date _____

MiniWeapon Used _____

DARTBOARD RULES

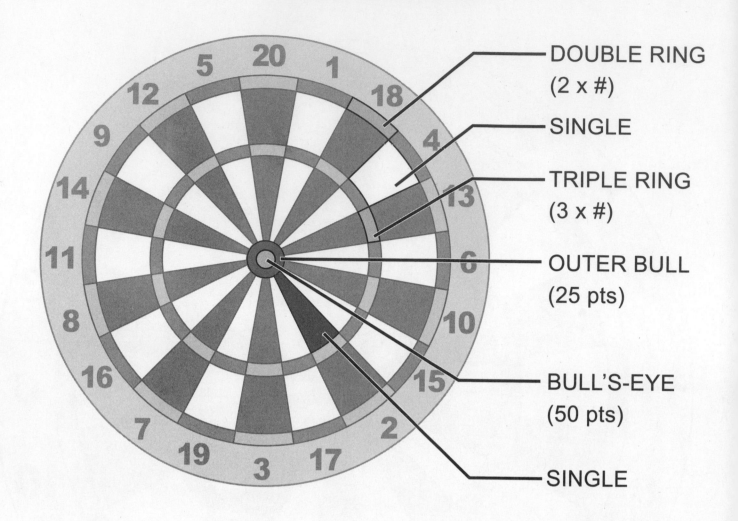

DOUBLE RING
(2 x #)

SINGLE

TRIPLE RING
(3 x #)

OUTER BULL
(25 pts)

BULL'S-EYE
(50 pts)

SINGLE

Basic Rules ★

Alternating turns, each player fires three times at the target from 5 to 12 feet away. When each player is finished shooting, add up his or her points and continue the game with the next player until someone has scored 101 or more. This is a slightly modified version of the traditional 501 or 301 dartboard game.

Marksman Rules ★★★

Alternating turns, each player fires three times at the dartboard from a distance of 9 feet. When each player is finished shooting, add up the points and continue the game with the next player until someone has scored 301 or more. For a different challenge with a simpler scoring system, use the chart printed underneath the dartboard. Each time a player hits a number between 15 and 20 or the bull's eye, fill in one of the corresponding circles. The first player to fill in all the circles wins.

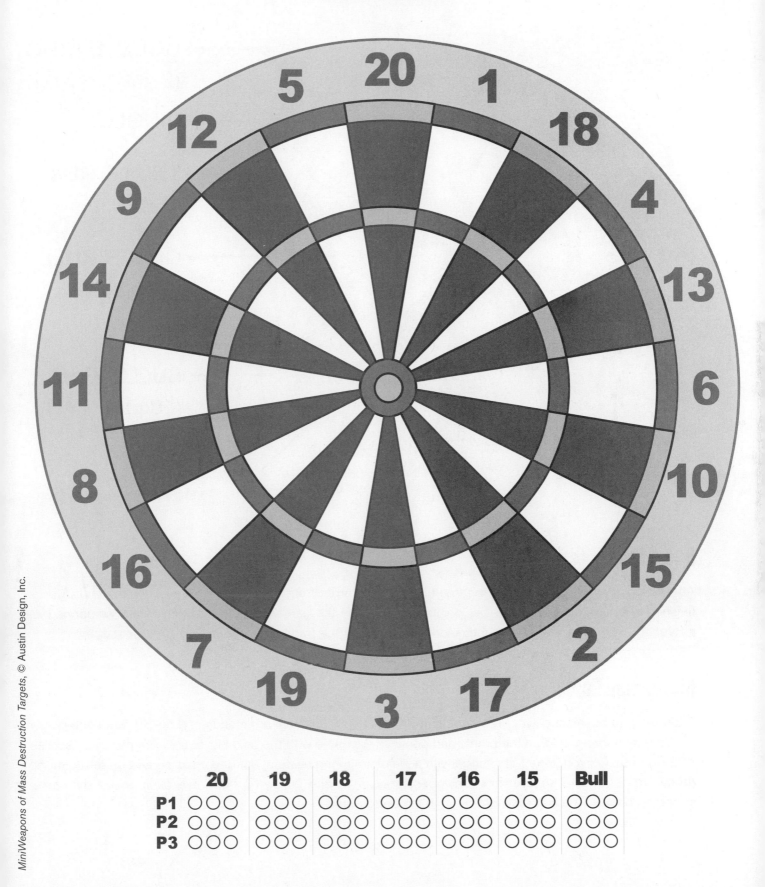

	20	19	18	17	16	15	Bull
P1	○○○	○○○	○○○	○○○	○○○	○○○	○○○
P2	○○○	○○○	○○○	○○○	○○○	○○○	○○○
P3	○○○	○○○	○○○	○○○	○○○	○○○	○○○

Competitor(s) _____ Date _____

MiniWeapon Used _____

DARTBOARD RULES

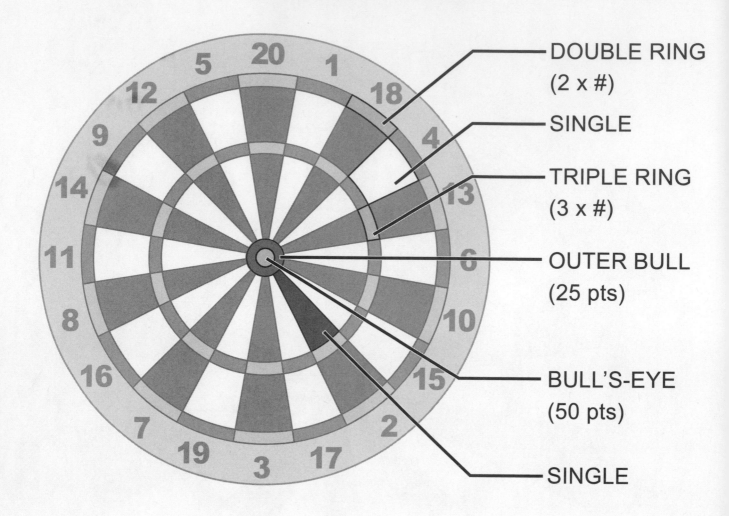

DOUBLE RING (2 x #)

SINGLE

TRIPLE RING (3 x #)

OUTER BULL (25 pts)

BULL'S-EYE (50 pts)

SINGLE

Basic Rules ★

Alternating turns, each player fires three times at the target from 5 to 12 feet away. When each player is finished shooting, add up his or her points and continue the game with the next player until someone has scored 101 or more. This is a slightly modified version of the traditional 501 or 301 dartboard game.

Marksman Rules ★★★

Alternating turns, each player fires three times at the dartboard from a distance of 9 feet. When each player is finished shooting, add up the points and continue the game with the next player until someone has scored 301 or more. For a different challenge with a simpler scoring system, use the chart printed underneath the dartboard. Each time a player hits a number between 15 and 20 or the bull's eye, fill in one of the corresponding circles. The first player to fill in all the circles wins.

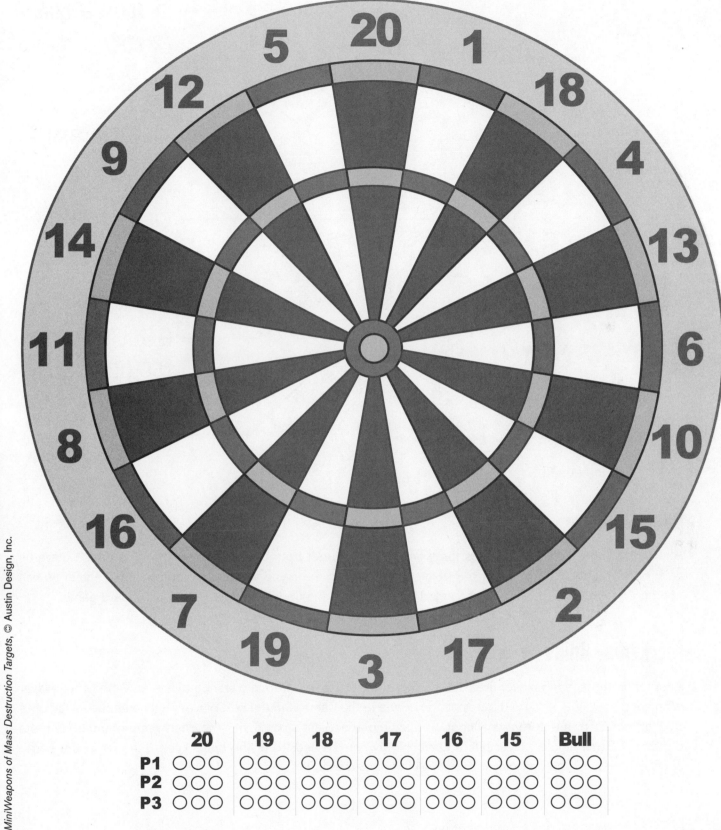

	20			19			18			17			16			15			Bull		
P1	○	○	○	○	○	○	○	○	○	○	○	○	○	○	○	○	○	○	○	○	○
P2	○	○	○	○	○	○	○	○	○	○	○	○	○	○	○	○	○	○	○	○	○
P3	○	○	○	○	○	○	○	○	○	○	○	○	○	○	○	○	○	○	○	○	○

Competitor(s) _____ Date _____

MiniWeapon Used _____

DARTBOARD RULES

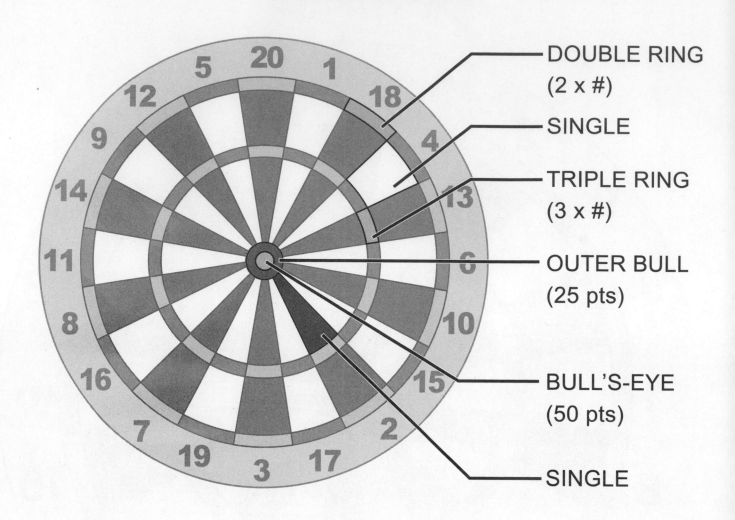

DOUBLE RING
(2 x #)

SINGLE

TRIPLE RING
(3 x #)

OUTER BULL
(25 pts)

BULL'S-EYE
(50 pts)

SINGLE

Basic Rules ★

Alternating turns, each player fires three times at the target from 5 to 12 feet away. When each player is finished shooting, add up his or her points and continue the game with the next player until someone has scored 101 or more. This is a slightly modified version of the traditional 501 or 301 dartboard game.

Marksman Rules ★★★

Alternating turns, each player fires three times at the dartboard from a distance of 9 feet. When each player is finished shooting, add up the points and continue the game with the next player until someone has scored 301 or more. For a different challenge with a simpler scoring system, use the chart printed underneath the dartboard. Each time a player hits a number between 15 and 20 or the bull's eye, fill in one of the corresponding circles. The first player to fill in all the circles wins.

	20	**19**	**18**	**17**	**16**	**15**	**Bull**
P1	○○○	○○○	○○○	○○○	○○○	○○○	○○○
P2	○○○	○○○	○○○	○○○	○○○	○○○	○○○
P3	○○○	○○○	○○○	○○○	○○○	○○○	○○○

Competitor(s) _____ Date _____

MiniWeapon Used _____

DARTBOARD RULES

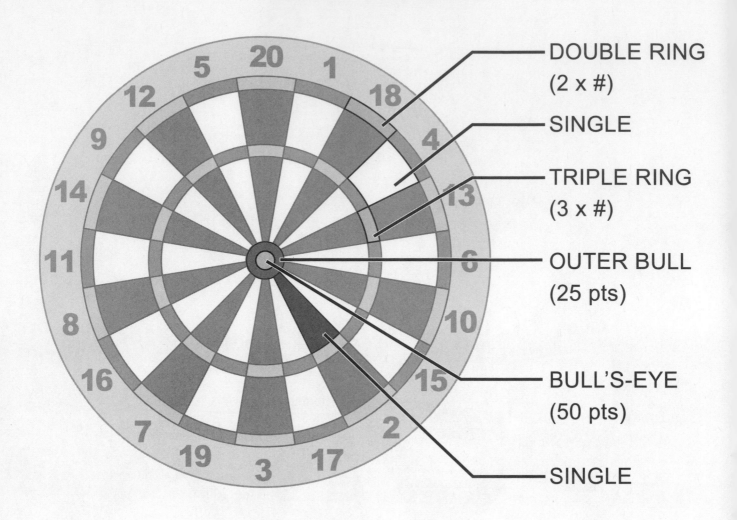

DOUBLE RING
(2 x #)

SINGLE

TRIPLE RING
(3 x #)

OUTER BULL
(25 pts)

BULL'S-EYE
(50 pts)

SINGLE

Basic Rules ★

Alternating turns, each player fires three times at the target from 5 to 12 feet away. When each player is finished shooting, add up his or her points and continue the game with the next player until someone has scored 101 or more. This is a slightly modified version of the traditional 501 or 301 dartboard game.

Marksman Rules ★★★

Alternating turns, each player fires three times at the dartboard from a distance of 9 feet. When each player is finished shooting, add up the points and continue the game with the next player until someone has scored 301 or more. For a different challenge with a simpler scoring system, use the chart printed underneath the dartboard. Each time a player hits a number between 15 and 20 or the bull's eye, fill in one of the corresponding circles. The first player to fill in all the circles wins.

	20			**19**			**18**			**17**			**16**			**15**			**Bull**		
P1	○	○	○	○	○	○	○	○	○	○	○	○	○	○	○	○	○	○	○	○	○
P2	○	○	○	○	○	○	○	○	○	○	○	○	○	○	○	○	○	○	○	○	○
P3	○	○	○	○	○	○	○	○	○	○	○	○	○	○	○	○	○	○	○	○	○

Competitor(s) _____   Date _____

MiniWeapon Used _____

DARTBOARD RULES

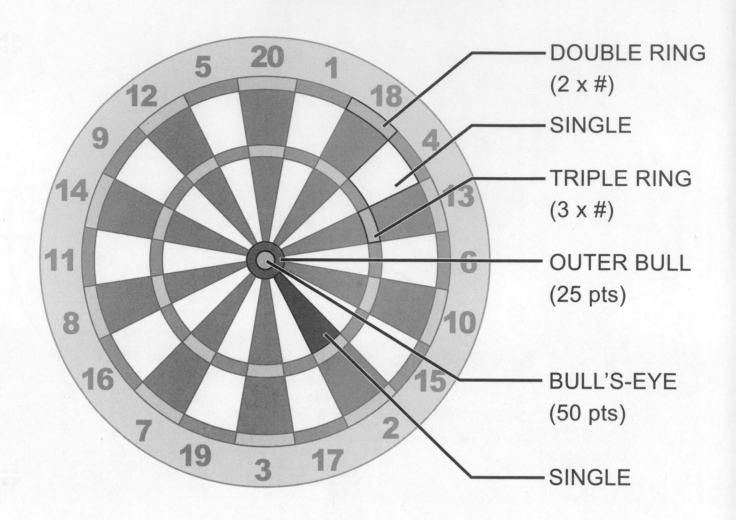

DOUBLE RING
(2 x #)

SINGLE

TRIPLE RING
(3 x #)

OUTER BULL
(25 pts)

BULL'S-EYE
(50 pts)

SINGLE

Basic Rules ★

Alternating turns, each player fires three times at the target from 5 to 12 feet away. When each player is finished shooting, add up his or her points and continue the game with the next player until someone has scored 101 or more. This is a slightly modified version of the traditional 501 or 301 dartboard game.

Marksman Rules ★★★

Alternating turns, each player fires three times at the dartboard from a distance of 9 feet. When each player is finished shooting, add up the points and continue the game with the next player until someone has scored 301 or more. For a different challenge with a simpler scoring system, use the chart printed underneath the dartboard. Each time a player hits a number between 15 and 20 or the bull's eye, fill in one of the corresponding circles. The first player to fill in all the circles wins.

MINI BASIC TARGETS

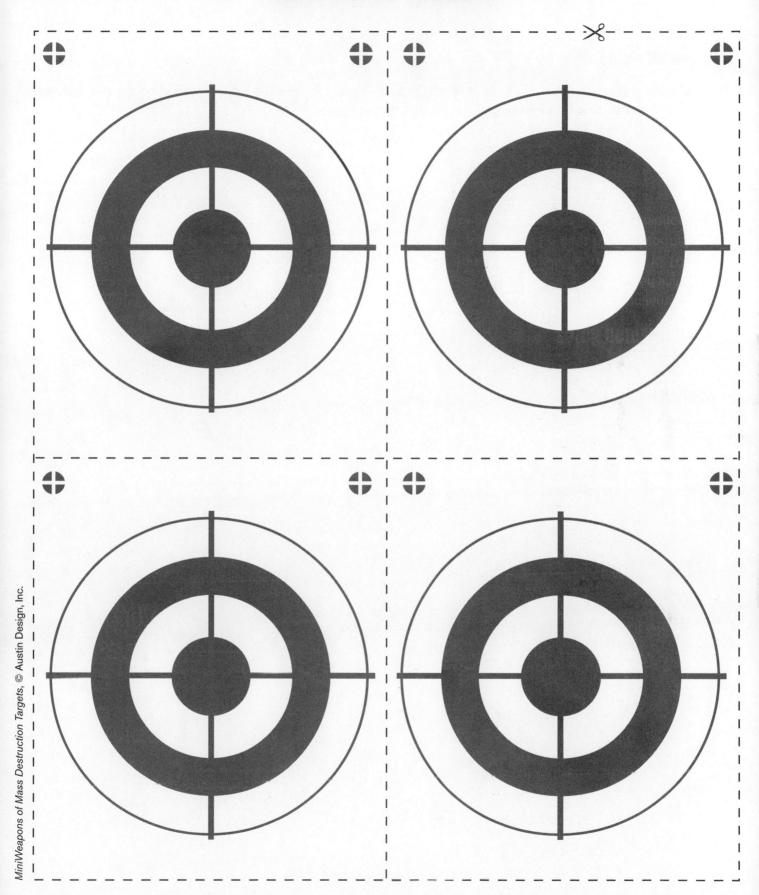

MINI BASIC TARGET RULES

Basic Rules ★

Once cut apart, the Mini Basic Targets are the perfect size for miniature arsenal testing on a wall or mounted on a tabletop. Use them for target practice or custom games.

MINI BASIC TARGETS

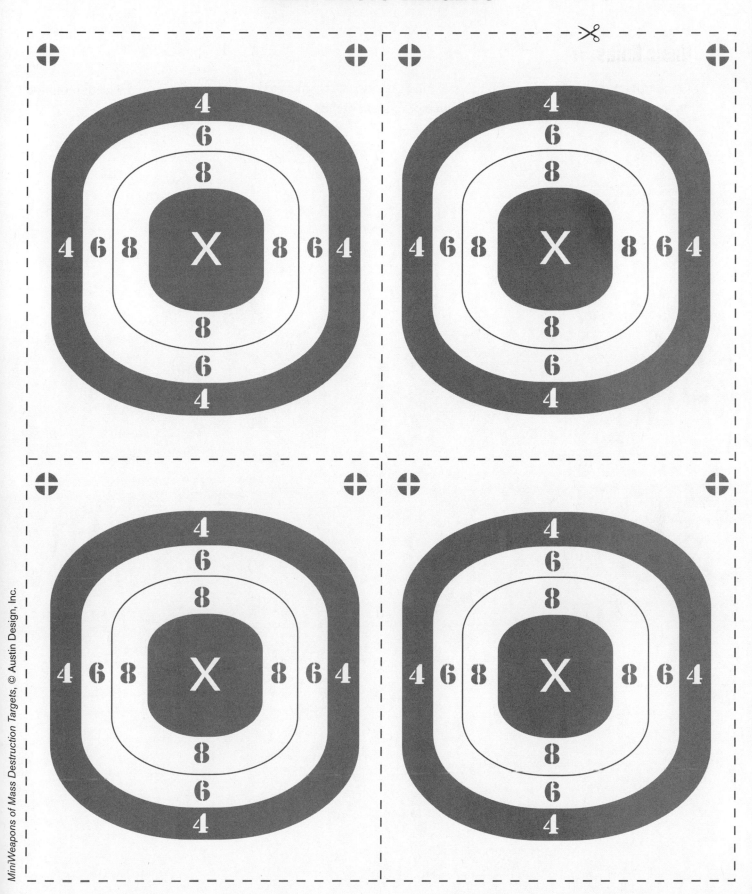

Basic Rules ★

Once cut apart, the Mini Basic Targets are the perfect size for miniature arsenal testing on a wall or mounted on a tabletop. Use them for target practice or custom games.

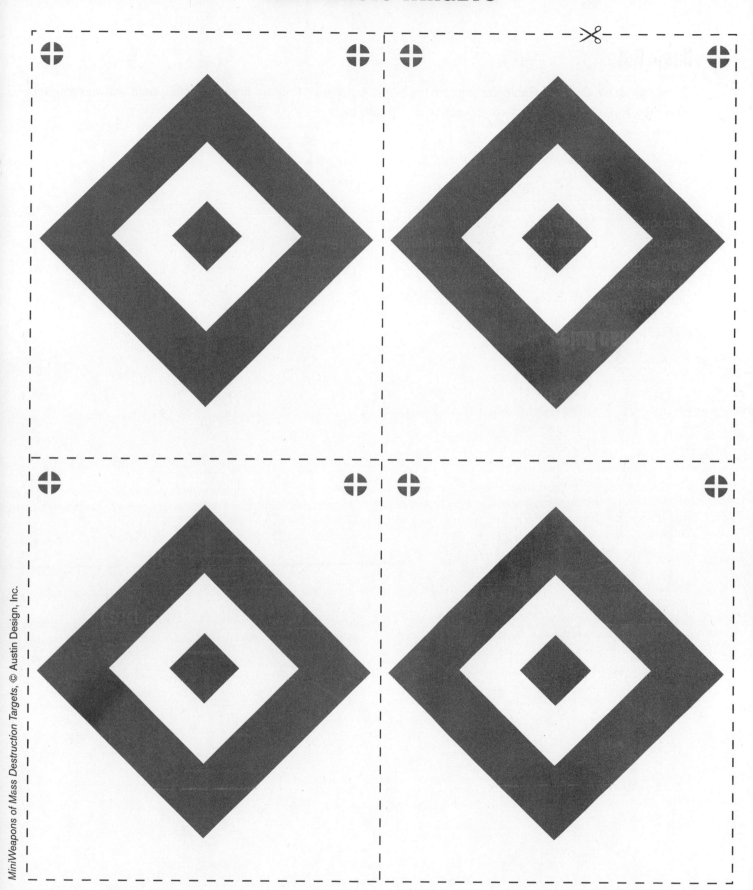

MINI BASIC TARGET RULES

Basic Rules ★

Once cut apart, the Mini Basic Targets are the perfect size for miniature arsenal testing on a wall or mounted on a tabletop. Use them for target practice or custom games.

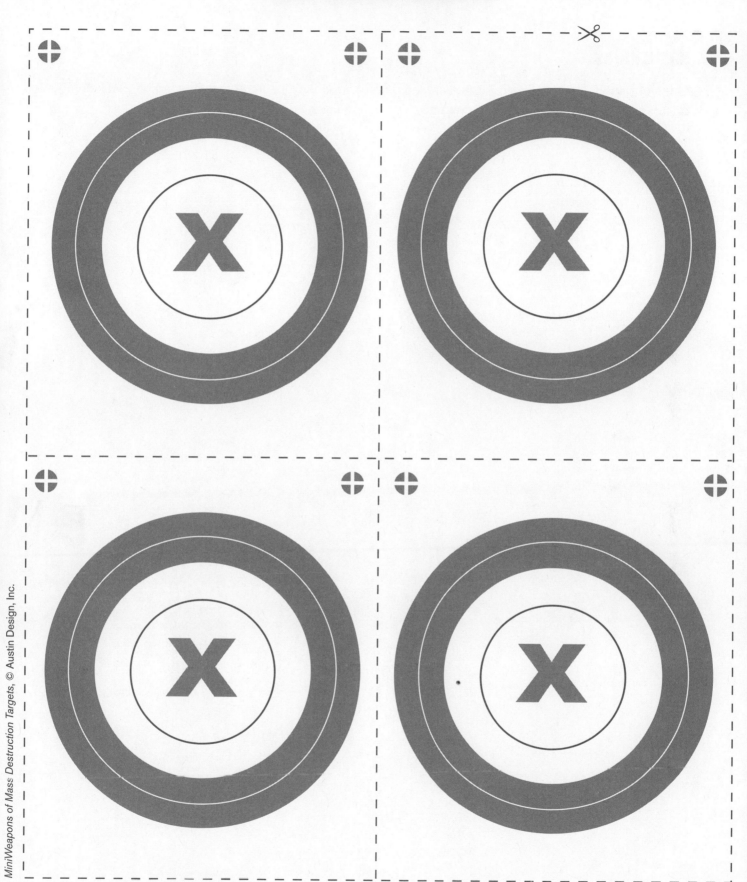

MINI BASIC TARGET RULES

Basic Rules ★

Once cut apart, the Mini Basic Targets are the perfect size for miniature arsenal testing on a wall or mounted on a tabletop. Use them for target practice or custom games.

SECRET AGENT TRAINING

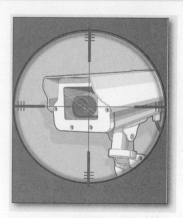

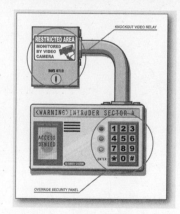

MiniWeapons of Mass Destruction Targets, © Austin Design, Inc.

Competitor(s) _____ Date _____

MiniWeapon Used _____

GUARD TOWER RULES

Basic Rules ★

To infiltrate the compound, you must first neutralize the rotating searchlight mounted on lookout tower 5. Security is tight, so you'll only have time for three quick shots. Once the searchlight is knocked out, move on! The first agent to complete this task wins.

Marksman Rules ★★★

Great shot—the searchlight has been destroyed. But the stationed guard still has communication capabilities. Take another three shots at the satellite dish fixed on the roof and/or the small telephone mounted to the right post. Either shot will silence the guards long enough for you to slip into the facility. The first agent to be successful is the winner.

Competitor(s) _____   Date _____

MiniWeapon Used _____

GUARD TOWER RULES

Basic Rules ★

To infiltrate the compound, you must first neutralize the rotating searchlight mounted on lookout tower 5. Security is tight, so you'll only have time for three quick shots. Once the searchlight is knocked out, move on! The first agent to complete this task wins.

Marksman Rules ★★★

Great shot—the searchlight has been destroyed. But the stationed guard still has communication capabilities. Take another three shots at the satellite dish fixed on the roof and/or the small telephone mounted to the right post. Either shot will silence the guards long enough for you to slip into the facility. The first agent to be successful is the winner.

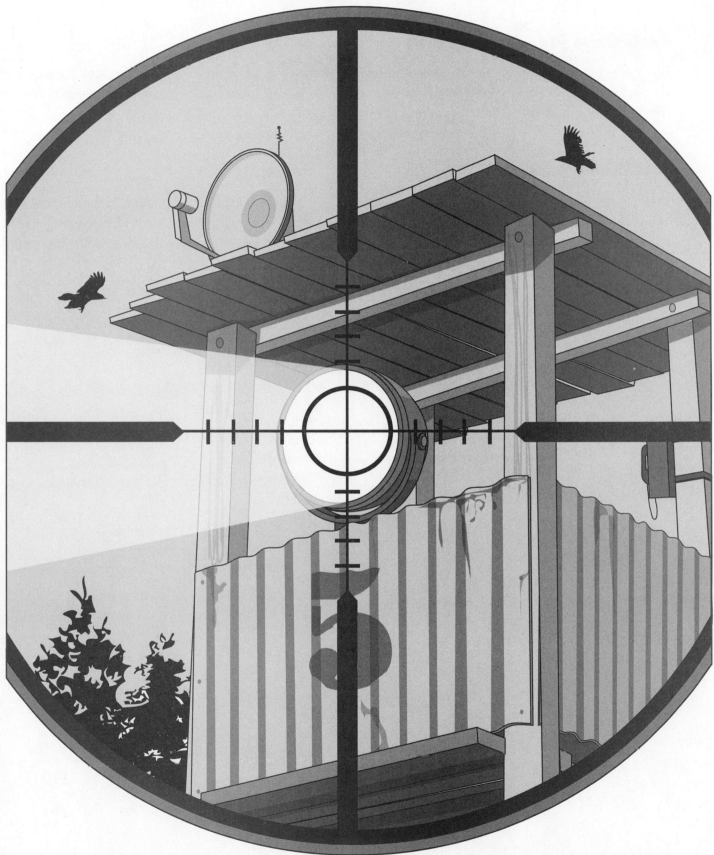

Competitor(s) _____ Date _____

MiniWeapon Used _____

GUARD TOWER RULES

Basic Rules ★

To infiltrate the compound, you must first neutralize the rotating searchlight mounted on lookout tower 5. Security is tight, so you'll only have time for three quick shots. Once the searchlight is knocked out, move on! The first agent to complete this task wins.

Marksman Rules ★ ★ ★

Great shot—the searchlight has been destroyed. But the stationed guard still has communication capabilities. Take another three shots at the satellite dish fixed on the roof and/or the small telephone mounted to the right post. Either shot will silence the guards long enough for you to slip into the facility. The first agent to be successful is the winner.

MiniWeapons of Mass Destruction Targets, © Austin Design, Inc.

Competitor(s) _____   Date _____

MiniWeapon Used _____

GUARD TOWER RULES

Basic Rules ★

To infiltrate the compound, you must first neutralize the rotating searchlight mounted on lookout tower 5. Security is tight, so you'll only have time for three quick shots. Once the searchlight is knocked out, move on! The first agent to complete this task wins.

Marksman Rules ★★★

Great shot—the searchlight has been destroyed. But the stationed guard still has communication capabilities. Take another three shots at the satellite dish fixed on the roof and/or the small telephone mounted to the right post. Either shot will silence the guards long enough for you to slip into the facility. The first agent to be successful is the winner.

MiniWeapons of Mass Destruction Targets. © Austin Design, Inc.

Competitor(s) _____ Date _____

MiniWeapon Used _____ **101**

GUARD TOWER RULES

Basic Rules ★

To infiltrate the compound, you must first neutralize the rotating searchlight mounted on lookout tower 5. Security is tight, so you'll only have time for three quick shots. Once the searchlight is knocked out, move on! The first agent to complete this task wins.

Marksman Rules ★★★

Great shot—the searchlight has been destroyed. But the stationed guard still has communication capabilities. Take another three shots at the satellite dish fixed on the roof and/or the small telephone mounted to the right post. Either shot will silence the guards long enough for you to slip into the facility. The first agent to be successful is the winner.

NOTICE
AUTHORIZED PERSONNEL ONLY

LOCK

Competitor(s) _____ Date _____

MiniWeapon Used _____

BREAK IN RULES

Basic Rules ★

With a tight schedule, picking this lock is not an option. Instead, blast it apart! With a direct hit, the chains will fall, giving you an invitation for intrusion. The first shooter to complete the task is the victor.

Marksman Rules ★★★

If the lock's armored housing is impenetrable, go for the chains. Unfortunately, it'll take an expert shooter to hit such a small target! The first infiltrator who makes the shot wins.

NOTICE

AUTHORIZED PERSONNEL ONLY

LOCK

Competitor(s) _____ Date _____

MiniWeapon Used _____

BREAK IN RULES

Basic Rules ★

With a tight schedule, picking this lock is not an option. Instead, blast it apart! With a direct hit, the chains will fall, giving you an invitation for intrusion. The first shooter to complete the task is the victor.

Marksman Rules ★★★

If the lock's armored housing is impenetrable, go for the chains. Unfortunately, it'll take an expert shooter to hit such a small target! The first infiltrator who makes the shot wins.

NOTICE

AUTHORIZED PERSONNEL ONLY

LOCK

Competitor(s) _____ Date _____

MiniWeapon Used _____

BREAK IN RULES

Basic Rules ★

With a tight schedule, picking this lock is not an option. Instead, blast it apart! With a direct hit, the chains will fall, giving you an invitation for intrusion. The first shooter to complete the task is the victor.

Marksman Rules ★ ★ ★

If the lock's armored housing is impenetrable, go for the chains. Unfortunately, it'll take an expert shooter to hit such a small target! The first infiltrator who makes the shot wins.

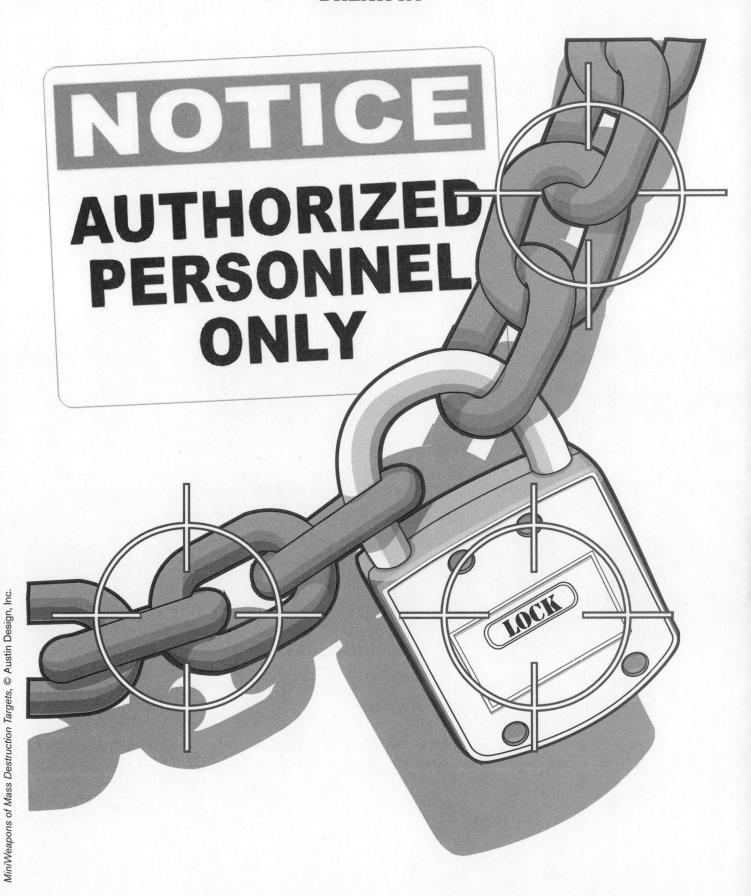

NOTICE

AUTHORIZED PERSONNEL ONLY

LOCK

Competitor(s) _____   Date _____

MiniWeapon Used _____

BREAK IN RULES

Basic Rules ★

With a tight schedule, picking this lock is not an option. Instead, blast it apart! With a direct hit, the chains will fall, giving you an invitation for intrusion. The first shooter to complete the task is the victor.

Marksman Rules ★★★

If the lock's armored housing is impenetrable, go for the chains. Unfortunately, it'll take an expert shooter to hit such a small target! The first infiltrator who makes the shot wins.

NOTICE

AUTHORIZED PERSONNEL ONLY

LOCK

Competitor(s) _____   Date _____

MiniWeapon Used _____

BREAK IN RULES

Basic Rules ★

With a tight schedule, picking this lock is not an option. Instead, blast it apart! With a direct hit, the chains will fall, giving you an invitation for intrusion. The first shooter to complete the task is the victor.

Marksman Rules ★★★

If the lock's armored housing is impenetrable, go for the chains. Unfortunately, it'll take an expert shooter to hit such a small target! The first infiltrator who makes the shot wins.

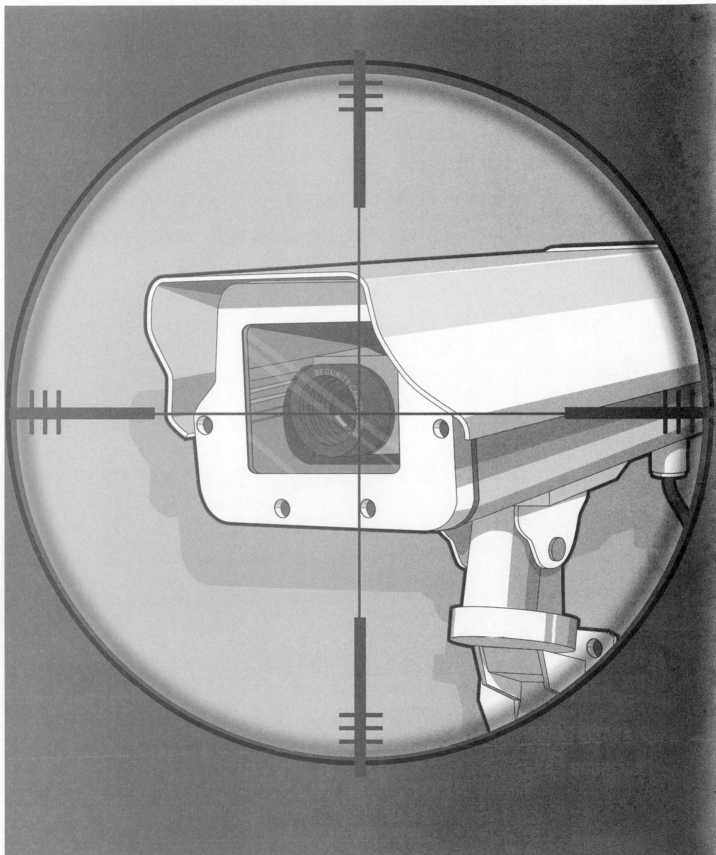

MiniWeapons of Mass Destruction Targets, © Austin Design, Inc.

Competitor(s) _____ Date _____

MiniWeapon Used _____

CAMERA KNOCKOUT RULES

Basic Rules ★

Stop—the corridor is being monitored by video surveillance! A direct hit should deactivate the camera. Place this target at least 10 feet away. A lens hit is an automatic win, but the camera's protective housing will need to be hit twice to knock it out. The first agent to complete either task claims the victory.

Marksman Rules ★ ★ ★

For real mission experience, mount one or two of these targets at an elevated height, and go for the lens. Modern, sophisticated security cameras are equipped with motion detectors and rotating capabilities, so you'll only have time for two shots before the control room identifies you. Shoot fast! The marksman who takes out the target first is the winner.

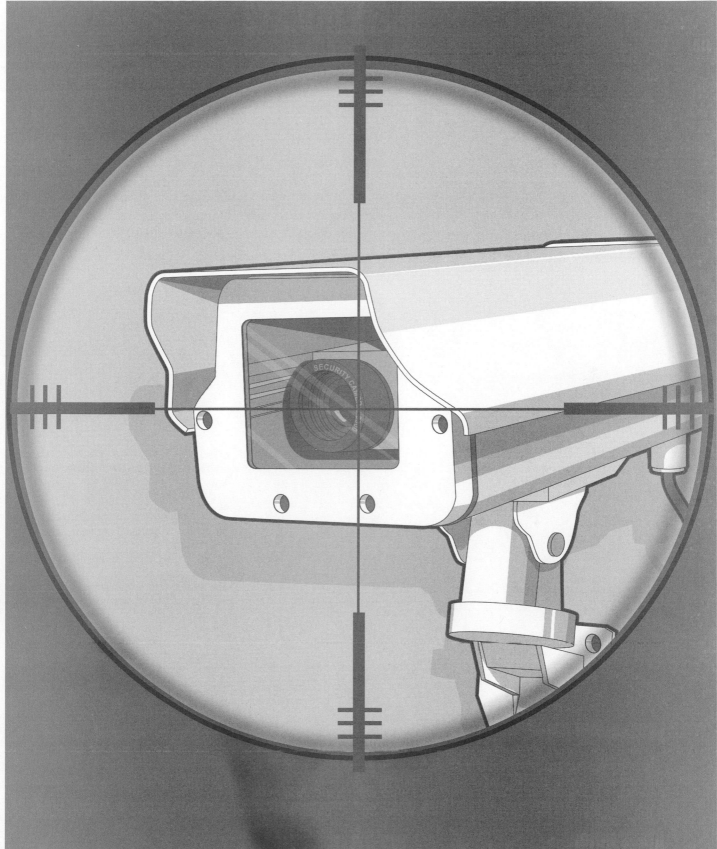

Competitor(s) _____ Date _____

MiniWeapon Used _____

CAMERA KNOCKOUT RULES

Basic Rules ★

Stop—the corridor is being monitored by video surveillance! A direct hit should deactivate the camera. Place this target at least 10 feet away. A lens hit is an automatic win, but the camera's protective housing will need to be hit twice to knock it out. The first agent to complete either task claims the victory.

Marksman Rules ★ ★ ★

For real mission experience, mount one or two of these targets at an elevated height, and go for the lens. Modern, sophisticated security cameras are equipped with motion detectors and rotating capabilities, so you'll only have time for two shots before the control room identifies you. Shoot fast! The marksman who takes out the target first is the winner.

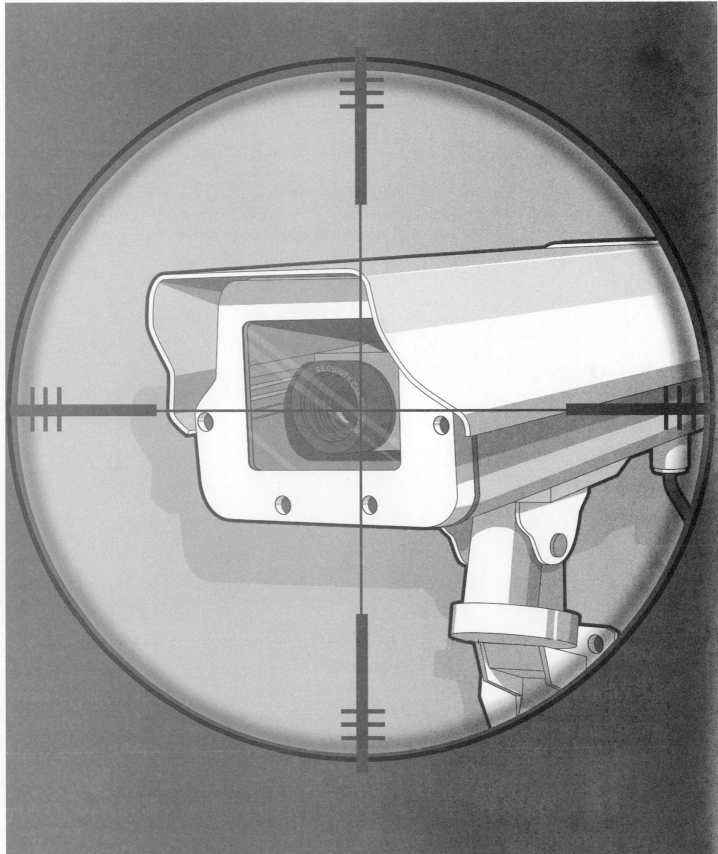

Competitor(s) _____ Date _____

MiniWeapon Used _____

CAMERA KNOCKOUT RULES

Basic Rules ★

Stop—the corridor is being monitored by video surveillance! A direct hit should deactivate the camera. Place this target at least 10 feet away. A lens hit is an automatic win, but the camera's protective housing will need to be hit twice to knock it out. The first agent to complete either task claims the victory.

Marksman Rules ★★★

For real mission experience, mount one or two of these targets at an elevated height, and go for the lens. Modern, sophisticated security cameras are equipped with motion detectors and rotating capabilities, so you'll only have time for two shots before the control room identifies you. Shoot fast! The marksman who takes out the target first is the winner.

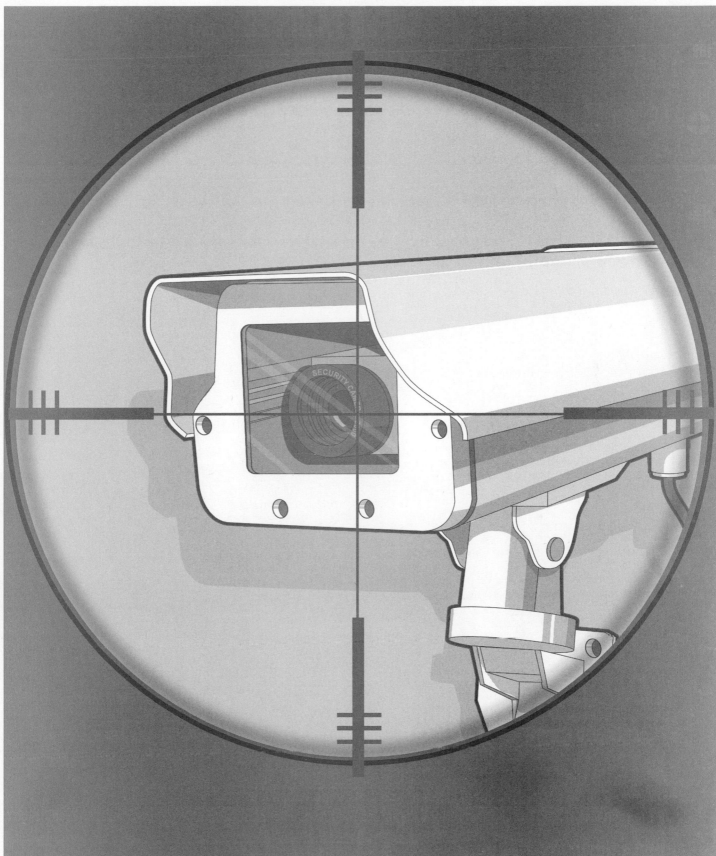

Competitor(s) _____ Date _____

MiniWeapon Used _____

CAMERA KNOCKOUT RULES

Basic Rules ★

Stop—the corridor is being monitored by video surveillance! A direct hit should deactivate the camera. Place this target at least 10 feet away. A lens hit is an automatic win, but the camera's protective housing will need to be hit twice to knock it out. The first agent to complete either task claims the victory.

Marksman Rules ★ ★ ★

For real mission experience, mount one or two of these targets at an elevated height, and go for the lens. Modern, sophisticated security cameras are equipped with motion detectors and rotating capabilities, so you'll only have time for two shots before the control room identifies you. Shoot fast! The marksman who takes out the target first is the winner.

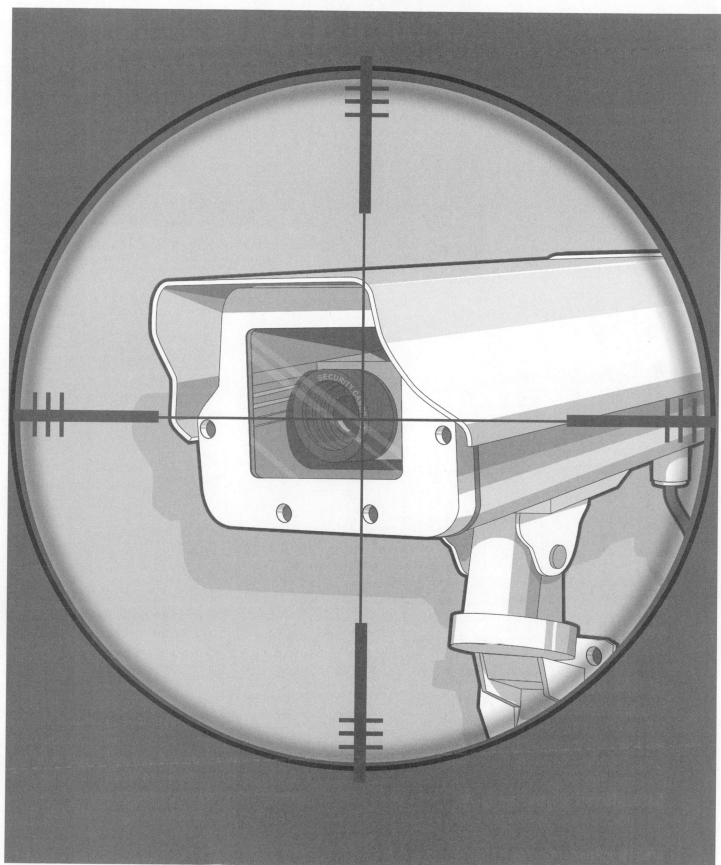

Competitor(s) _____ Date _____

MiniWeapon Used _____

CAMERA KNOCKOUT RULES

Basic Rules ★

Stop—the corridor is being monitored by video surveillance! A direct hit should deactivate the camera. Place this target at least 10 feet away. A lens hit is an automatic win, but the camera's protective housing will need to be hit twice to knock it out. The first agent to complete either task claims the victory.

Marksman Rules ★ ★ ★

For real mission experience, mount one or two of these targets at an elevated height, and go for the lens. Modern, sophisticated security cameras are equipped with motion detectors and rotating capabilities, so you'll only have time for two shots before the control room identifies you. Shoot fast! The marksman who takes out the target first is the winner.

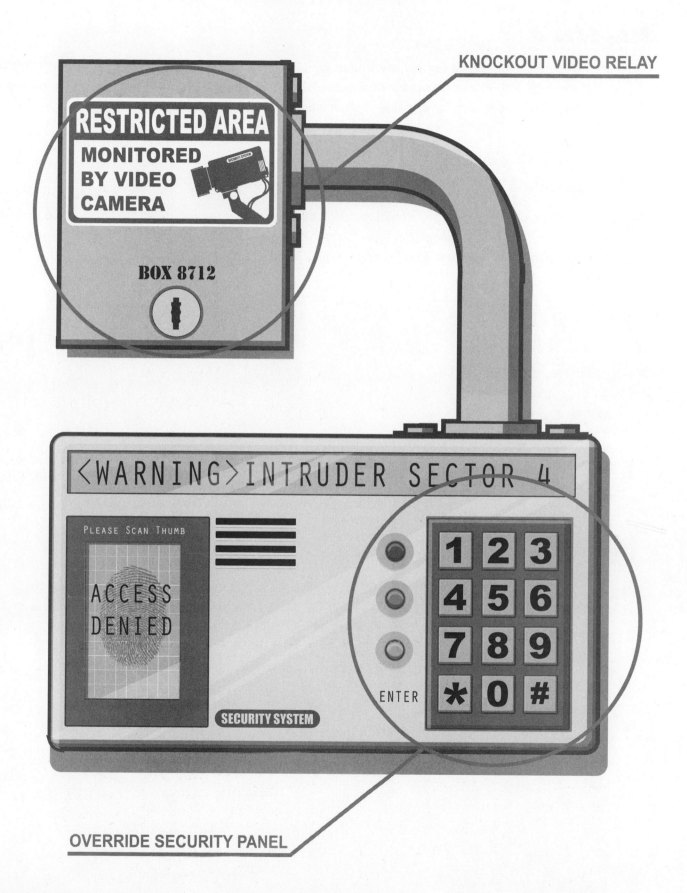

KNOCKOUT VIDEO RELAY

RESTRICTED AREA
MONITORED
BY VIDEO
CAMERA

BOX 8712

<WARNING>INTRUDER SECTOR 4

PLEASE SCAN THUMB

ACCESS
DENIED

SECURITY SYSTEM

ENTER

1 2 3
4 5 6
7 8 9
* 0 #

OVERRIDE SECURITY PANEL

Competitor(s) _____   Date _____

MiniWeapon Used _____

DISABLE ALARM RULES

Basic Rules ★

Rats! This climate-controlled laboratory is secured by a computerized fingerprint scanner! Good thing you did your homework and reviewed the electrical schematics prior to the mission. With a direct hit to the keypad, the scanner should short-circuit, giving you access. Complete the task before your fellow agents to win the challenge.

Marksman Rules ★★★

Knocking out the biometric keypad is a must, but as a secondary precaution you should deactivate the video relay box wired to the security panel. One quick shot should do it! Before firing, step back 10 feet to avoid the electrical explosion. The agent who takes it out first is the victor.

KNOCKOUT VIDEO RELAY

RESTRICTED AREA
MONITORED BY VIDEO CAMERA

BOX 8712

<WARNING>INTRUDER SECTOR 4

PLEASE SCAN THUMB

ACCESS DENIED

SECURITY SYSTEM

ENTER

OVERRIDE SECURITY PANEL

MiniWeapons of Mass Destruction Targets, © Austin Design, Inc.

Competitor(s) _____   Date _____

MiniWeapon Used _____

DISABLE ALARM RULES

Basic Rules ★

Rats! This climate-controlled laboratory is secured by a computerized fingerprint scanner! Good thing you did your homework and reviewed the electrical schematics prior to the mission. With a direct hit to the keypad, the scanner should short-circuit, giving you access. Complete the task before your fellow agents to win the challenge.

Marksman Rules ★★★

Knocking out the biometric keypad is a must, but as a secondary precaution you should deactivate the video relay box wired to the security panel. One quick shot should do it! Before firing, step back 10 feet to avoid the electrical explosion. The agent who takes it out first is the victor.

KNOCKOUT VIDEO RELAY

RESTRICTED AREA
MONITORED BY VIDEO CAMERA

SECURITY SYSTEM

BOX 8712

<WARNING> INTRUDER SECTOR 4

PLEASE SCAN THUMB

ACCESS DENIED

1 2 3
4 5 6
7 8 9
* 0 #

ENTER

SECURITY SYSTEM

OVERRIDE SECURITY PANEL

MiniWeapons of Mass Destruction Targets, © Austin Design, Inc.

Competitor(s) _____ Date _____

MiniWeapon Used _____

DISABLE ALARM RULES

Basic Rules ★

Rats! This climate-controlled laboratory is secured by a computerized fingerprint scanner! Good thing you did your homework and reviewed the electrical schematics prior to the mission. With a direct hit to the keypad, the scanner should short-circuit, giving you access. Complete the task before your fellow agents to win the challenge.

Marksman Rules ★★★

Knocking out the biometric keypad is a must, but as a secondary precaution you should deactivate the video relay box wired to the security panel. One quick shot should do it! Before firing, step back 10 feet to avoid the electrical explosion. The agent who takes it out first is the victor.

KNOCKOUT VIDEO RELAY

RESTRICTED AREA

MONITORED BY VIDEO CAMERA

BOX 8712

〈WARNING〉INTRUDER SECTOR 4

PLEASE SCAN THUMB

ACCESS DENIED

SECURITY SYSTEM

ENTER

1	2	3
4	5	6
7	8	9
*	0	#

OVERRIDE SECURITY PANEL

Competitor(s) _____   Date _____

MiniWeapon Used _____

DISABLE ALARM RULES

Basic Rules ★

Rats! This climate-controlled laboratory is secured by a computerized fingerprint scanner! Good thing you did your homework and reviewed the electrical schematics prior to the mission. With a direct hit to the keypad, the scanner should short-circuit, giving you access. Complete the task before your fellow agents to win the challenge.

Marksman Rules ★★★

Knocking out the biometric keypad is a must, but as a secondary precaution you should deactivate the video relay box wired to the security panel. One quick shot should do it! Before firing, step back 10 feet to avoid the electrical explosion. The agent who takes it out first is the victor.

KNOCKOUT VIDEO RELAY

RESTRICTED AREA
MONITORED
BY VIDEO
CAMERA

BOX 8712

<WARNING>INTRUDER SECTOR 4

PLEASE SCAN THUMB

ACCESS DENIED

SECURITY SYSTEM

ENTER

1 2 3
4 5 6
7 8 9
* 0 #

OVERRIDE SECURITY PANEL

MiniWeapons of Mass Destruction Targets, © Austin Design, Inc.

Competitor(s) _____ Date _____

MiniWeapon Used _____

DISABLE ALARM RULES

Basic Rules ★

Rats! This climate-controlled laboratory is secured by a computerized fingerprint scanner! Good thing you did your homework and reviewed the electrical schematics prior to the mission. With a direct hit to the keypad, the scanner should short-circuit, giving you access. Complete the task before your fellow agents to win the challenge.

Marksman Rules ★ ★ ★

Knocking out the biometric keypad is a must, but as a secondary precaution you should deactivate the video relay box wired to the security panel. One quick shot should do it! Before firing, step back 10 feet to avoid the electrical explosion. The agent who takes it out first is the victor.

Competitor(s) _____ Date _____

MiniWeapon Used _____

EXPLOSIVES RULES

Basic Rules ★

This laboratory must be destroyed! Everything is going according to plan until you notice the digital count-down has malfunctioned. As you hear the faint sound of footsteps, you realize that only a precision shot will detonate the explosives. Step back at least 8 feet and try to hit the bundle while avoiding the digital timer. This is dangerous work, but secret agent training has prepared you for situations like this. Victory goes to the first shooter to make the shot.

Marksman Rules ★★★

Stop! Escape Plan A has failed, and more time is needed for Plan B—an escape up the ventilation vent. With time ticking down, the bomb will need to be defused from a safe distance. Aim your MiniWeapon *at the digital timer only*, then fire. A direct hit should buy you more time to unscrew and remove the vent cover; just avoid targeting the wrapped explosives. The first agent to complete a direct hit wins.

Competitor(s) _____ Date _____

MiniWeapon Used _____

EXPLOSIVES RULES

Basic Rules ★

This laboratory must be destroyed! Everything is going according to plan until you notice the digital countdown has malfunctioned. As you hear the faint sound of footsteps, you realize that only a precision shot will detonate the explosives. Step back at least 8 feet and try to hit the bundle while avoiding the digital timer. This is dangerous work, but secret agent training has prepared you for situations like this. Victory goes to the first shooter to make the shot.

Marksman Rules ★★★

Stop! Escape Plan A has failed, and more time is needed for Plan B—an escape up the ventilation vent. With time ticking down, the bomb will need to be defused from a safe distance. Aim your MiniWeapon *at the digital timer only*, then fire. A direct hit should buy you more time to unscrew and remove the vent cover; just avoid targeting the wrapped explosives. The first agent to complete a direct hit wins.

MiniWeapons of Mass Destruction Targets, © Austin Design, Inc.

Competitor(s) _____ Date _____

MiniWeapon Used _____

EXPLOSIVES RULES

Basic Rules ★

This laboratory must be destroyed! Everything is going according to plan until you notice the digital count-down has malfunctioned. As you hear the faint sound of footsteps, you realize that only a precision shot will detonate the explosives. Step back at least 8 feet and try to hit the bundle while avoiding the digital timer. This is dangerous work, but secret agent training has prepared you for situations like this. Victory goes to the first shooter to make the shot.

Marksman Rules ★★★

Stop! Escape Plan A has failed, and more time is needed for Plan B—an escape up the ventilation vent. With time ticking down, the bomb will need to be defused from a safe distance. Aim your MiniWeapon *at the digital timer only*, then fire. A direct hit should buy you more time to unscrew and remove the vent cover; just avoid targeting the wrapped explosives. The first agent to complete a direct hit wins.

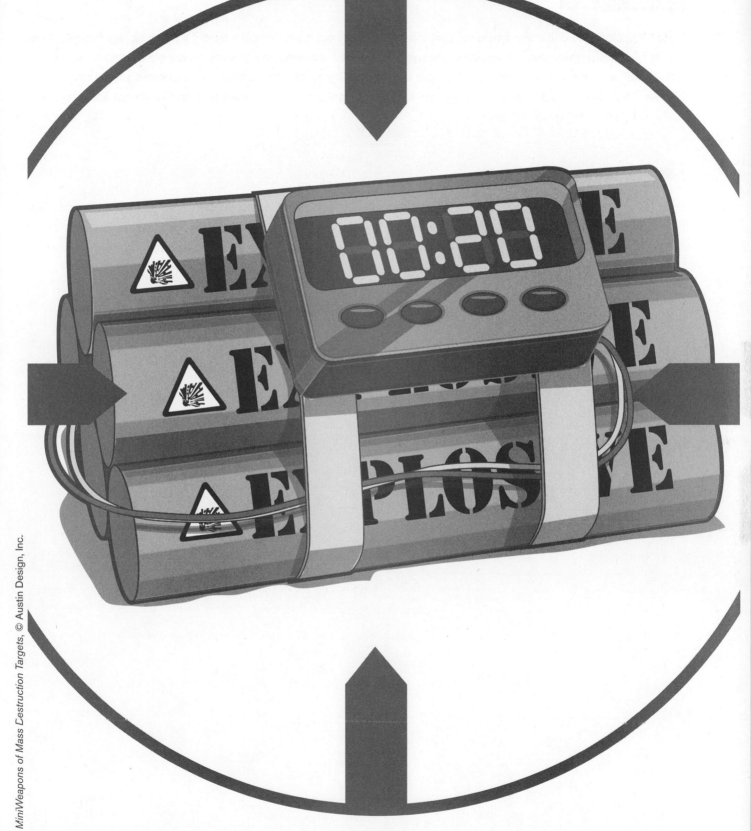

Competitor(s) _____ Date _____

MiniWeapon Used _____

EXPLOSIVES RULES

Basic Rules ★

This laboratory must be destroyed! Everything is going according to plan until you notice the digital countdown has malfunctioned. As you hear the faint sound of footsteps, you realize that only a precision shot will detonate the explosives. Step back at least 8 feet and try to hit the bundle while avoiding the digital timer. This is dangerous work, but secret agent training has prepared you for situations like this. Victory goes to the first shooter to make the shot.

Marksman Rules ★ ★ ★

Stop! Escape Plan A has failed, and more time is needed for Plan B—an escape up the ventilation vent. With time ticking down, the bomb will need to be defused from a safe distance. Aim your MiniWeapon *at the digital timer only*, then fire. A direct hit should buy you more time to unscrew and remove the vent cover; just avoid targeting the wrapped explosives. The first agent to complete a direct hit wins.

Competitor(s) _____ Date _____

MiniWeapon Used _____

EXPLOSIVES RULES

Basic Rules ★

This laboratory must be destroyed! Everything is going according to plan until you notice the digital count-down has malfunctioned. As you hear the faint sound of footsteps, you realize that only a precision shot will detonate the explosives. Step back at least 8 feet and try to hit the bundle while avoiding the digital timer. This is dangerous work, but secret agent training has prepared you for situations like this. Victory goes to the first shooter to make the shot.

Marksman Rules ★★★

Stop! Escape Plan A has failed, and more time is needed for Plan B—an escape up the ventilation vent. With time ticking down, the bomb will need to be defused from a safe distance. Aim your MiniWeapon *at the digital timer only*, then fire. A direct hit should buy you more time to unscrew and remove the vent cover; just avoid targeting the wrapped explosives. The first agent to complete a direct hit wins.

Competitor(s) _____   Date _____

MiniWeapon Used _____

LIGHTS OUT RULES

Basic Rules ★

Someone's in the room. Quick, knock out the lights! Take three quick shots to hit the bulb or its metal socket—either hit will do! Avoid the shattering glass by firing from a distance of 8 feet or more. The agent who completes this task first is the winner.

Marksman Rules ★ ★ ★

Looks like this room has multiple lights! Pin up two targets and try to nail both lights in two shots, firing from a distance of at least 10 feet. Be the first to hit both shots in a single turn to win the challenge.

Competitor(s) _____ Date _____

MiniWeapon Used _____

LIGHTS OUT RULES

Basic Rules ★

Someone's in the room. Quick, knock out the lights! Take three quick shots to hit the bulb or its metal socket—either hit will do! Avoid the shattering glass by firing from a distance of 8 feet or more. The agent who completes this task first is the winner.

Marksman Rules ★ ★ ★

Looks like this room has multiple lights! Pin up two targets and try to nail both lights in two shots, firing from a distance of at least 10 feet. Be the first to hit both shots in a single turn to win the challenge.

Competitor(s) _____ Date _____

MiniWeapon Used _____

LIGHTS OUT RULES

Basic Rules ★

Someone's in the room. Quick, knock out the lights! Take three quick shots to hit the bulb or its metal socket—either hit will do! Avoid the shattering glass by firing from a distance of 8 feet or more. The agent who completes this task first is the winner.

Marksman Rules ★ ★ ★

Looks like this room has multiple lights! Pin up two targets and try to nail both lights in two shots, firing from a distance of at least 10 feet. Be the first to hit both shots in a single turn to win the challenge.

Competitor(s) _____ Date _____

MiniWeapon Used _____

LIGHTS OUT RULES

Basic Rules ★

Someone's in the room. Quick, knock out the lights! Take three quick shots to hit the bulb or its metal socket—either hit will do! Avoid the shattering glass by firing from a distance of 8 feet or more. The agent who completes this task first is the winner.

Marksman Rules ★ ★ ★

Looks like this room has multiple lights! Pin up two targets and try to nail both lights in two shots, firing from a distance of at least 10 feet. Be the first to hit both shots in a single turn to win the challenge.

Competitor(s) _____ Date _____

MiniWeapon Used _____

LIGHTS OUT RULES

Basic Rules ★

Someone's in the room. Quick, knock out the lights! Take three quick shots to hit the bulb or its metal socket—either hit will do! Avoid the shattering glass by firing from a distance of 8 feet or more. The agent who completes this task first is the winner.

Marksman Rules ★ ★ ★

Looks like this room has multiple lights! Pin up two targets and try to nail both lights in two shots, firing from a distance of at least 10 feet. Be the first to hit both shots in a single turn to win the challenge.

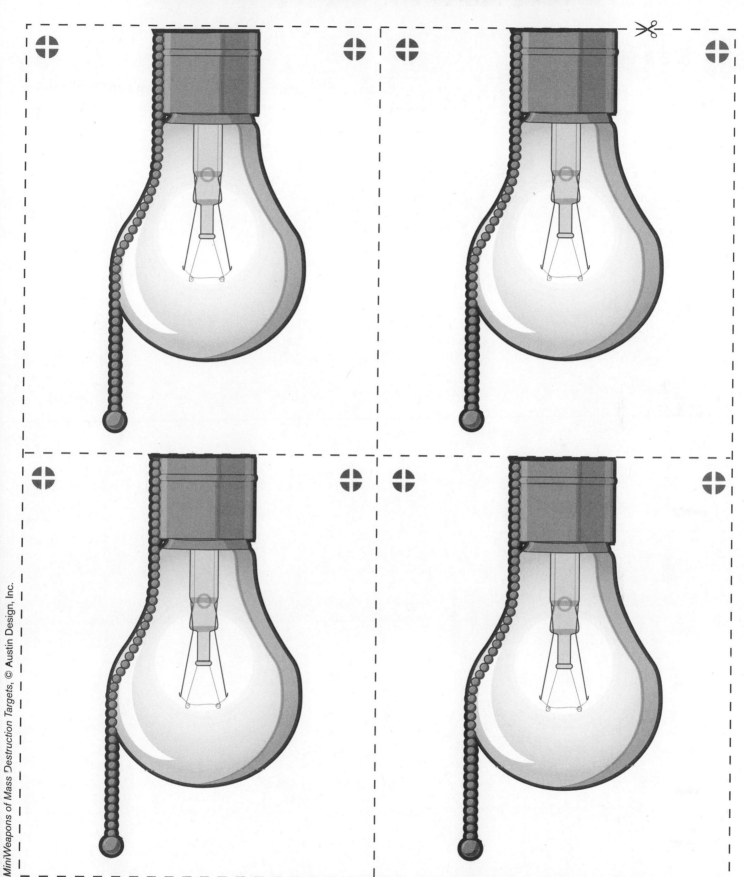

MINI SECRET AGENT TRAINING TARGETS RULES

Basic Rules ★

Once cut apart, the Mini Secret Agent Training Targets are the perfect size for miniature arsenal testing on a wall or mounted on a tabletop. Use them for target practice or custom games.

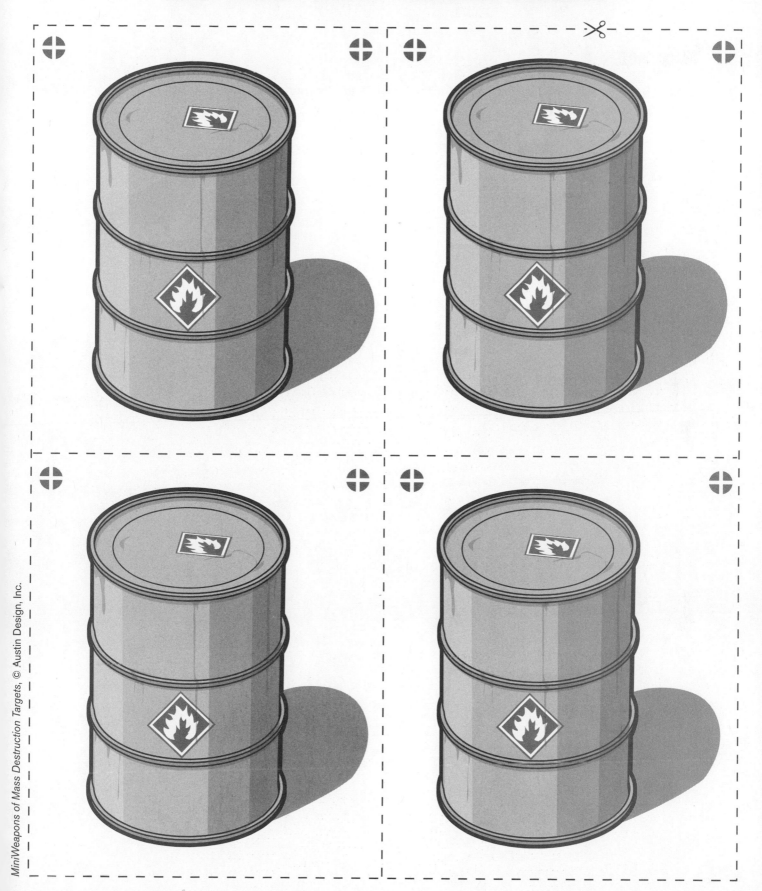

MINI SECRET AGENT TRAINING TARGETS RULES

Basic Rules ★

Once cut apart, the Mini Secret Agent Training Targets are the perfect size for miniature arsenal testing on a wall or mounted on a tabletop. Use them for target practice or custom games.

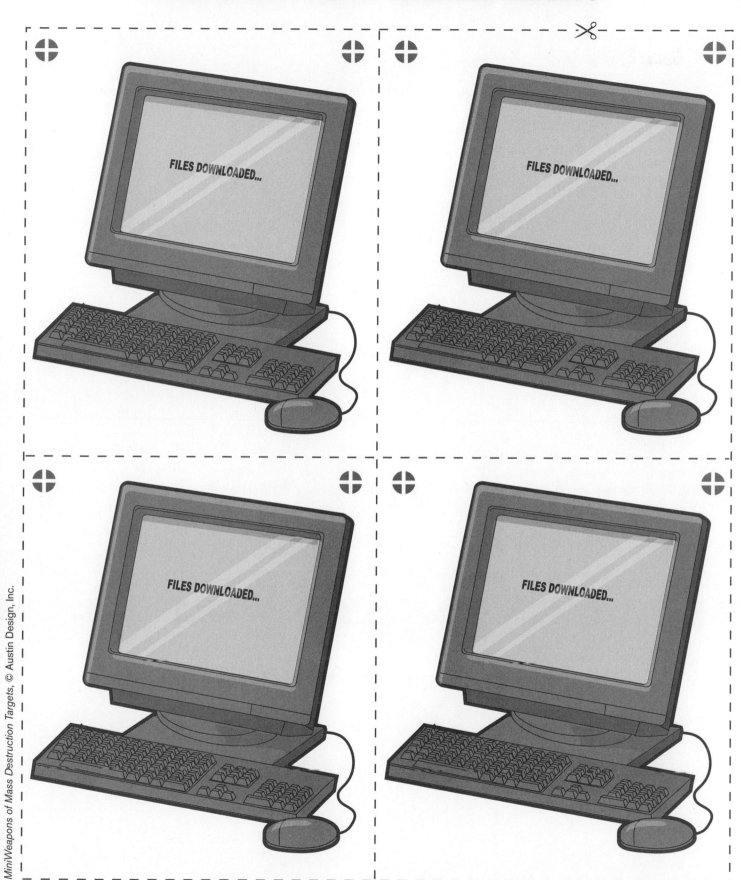

MINI SECRET AGENT TRAINING TARGETS RULES

Basic Rules ★

Once cut apart, the Mini Secret Agent Training Targets are the perfect size for miniature arsenal testing on a wall or mounted on a tabletop. Use them for target practice or custom games.

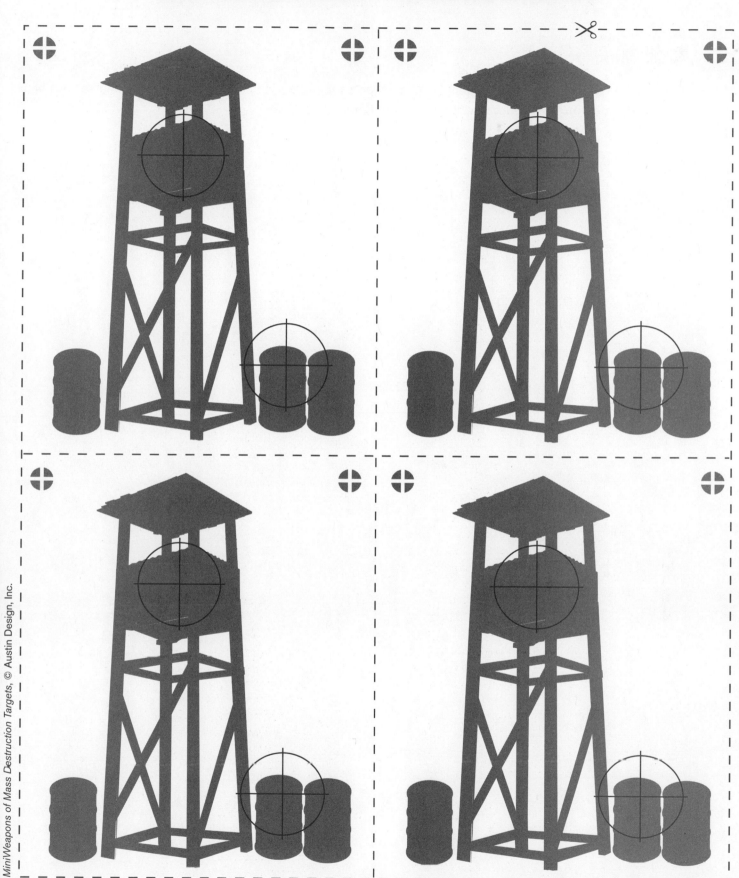

MiniWeapons of Mass Destruction Targets, © Austin Design, Inc.

MINI SECRET AGENT TRAINING TARGETS RULES

Basic Rules ★

Once cut apart, the Mini Secret Agent Training Targets are the perfect size for miniature arsenal testing on a wall or mounted on a tabletop. Use them for target practice or custom games.

DARK AGES

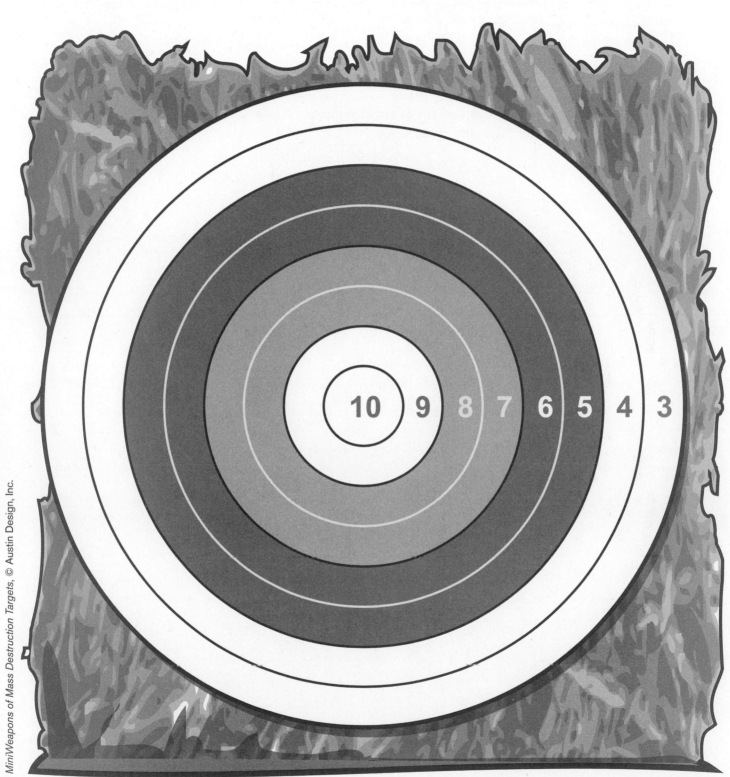

10 9 8 7 6 5 4 3

MiniWeapons of Mass Destruction Targets, © Austin Design, Inc.

Competitor(s) _____ Date _____

MiniWeapon Used _____

ARCHERY TARGET RULES

Basic Rules ★

Alternating turns, each player fires three or six shots at the target from 5 to 12 feet away. When each player is finished shooting, competitors add up the score; play continues until everyone has had a turn. The shooter with the highest total score wins.

Marksman Rules ★★★

Alternating turns, each player fires three or six shots at the target from at least 10 feet away. The first marksman to score a cumulative 50 points or higher wins!

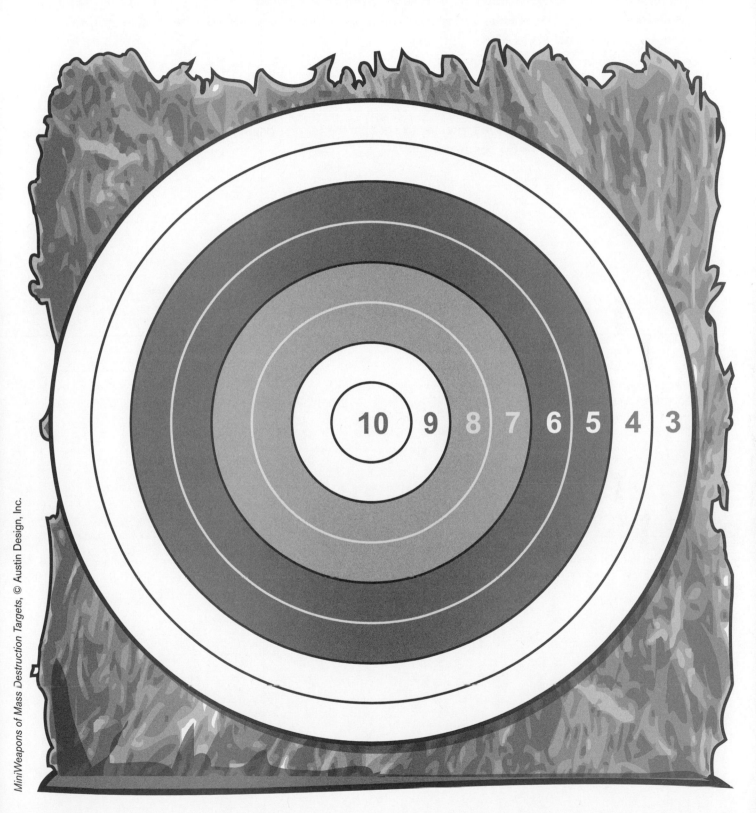

10 9 8 7 6 5 4 3

MiniWeapons of Mass Destruction Targets, © Austin Design, Inc.

Competitor(s) _____ Date _____

MiniWeapon Used _____

ARCHERY TARGET RULES

Basic Rules ★

Alternating turns, each player fires three or six shots at the target from 5 to 12 feet away. When each player is finished shooting, competitors add up the score; play continues until everyone has had a turn. The shooter with the highest total score wins.

Marksman Rules ★ ★ ★

Alternating turns, each player fires three or six shots at the target from at least 10 feet away. The first marksman to score a cumulative 50 points or higher wins!

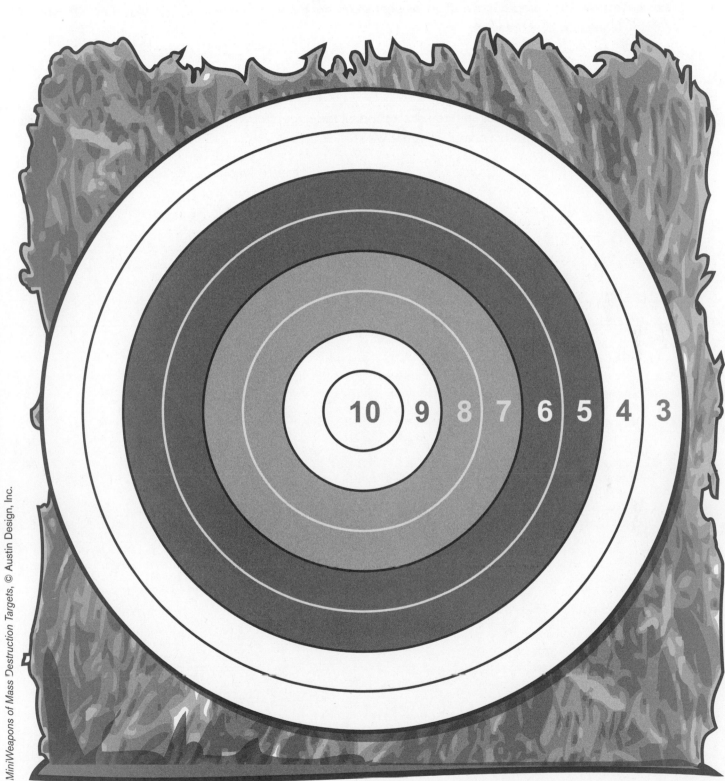

10 9 8 7 6 5 4 3

MiniWeapons of Mass Destruction Targets, © Austin Design, Inc.

Competitor(s) _____ Date _____

MiniWeapon Used _____

ARCHERY TARGET RULES

Basic Rules ★

Alternating turns, each player fires three or six shots at the target from 5 to 12 feet away. When each player is finished shooting, competitors add up the score; play continues until everyone has had a turn. The shooter with the highest total score wins.

Marksman Rules ★ ★ ★

Alternating turns, each player fires three or six shots at the target from at least 10 feet away. The first marksman to score a cumulative 50 points or higher wins!

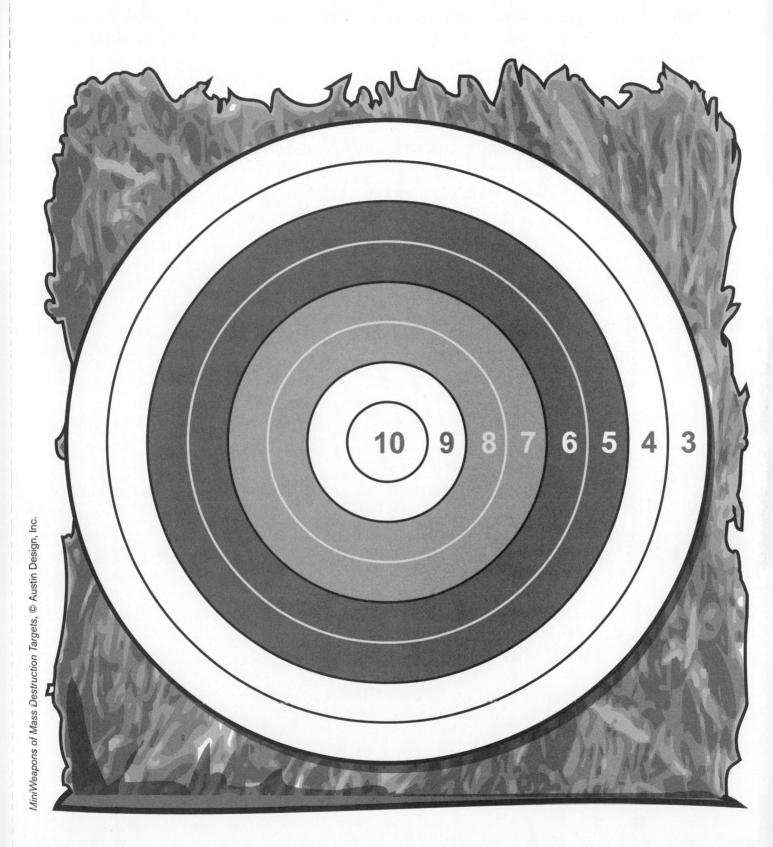

10 9 8 7 6 5 4 3

MiniWeapons of Mass Destruction Targets, © Austin Design, Inc.

Competitor(s) _____   Date _____

MiniWeapon Used _____

ARCHERY TARGET RULES

Basic Rules ★

Alternating turns, each player fires three or six shots at the target from 5 to 12 feet away. When each player is finished shooting, competitors add up the score; play continues until everyone has had a turn. The shooter with the highest total score wins.

Marksman Rules ★ ★ ★

Alternating turns, each player fires three or six shots at the target from at least 10 feet away. The first marksman to score a cumulative 50 points or higher wins!

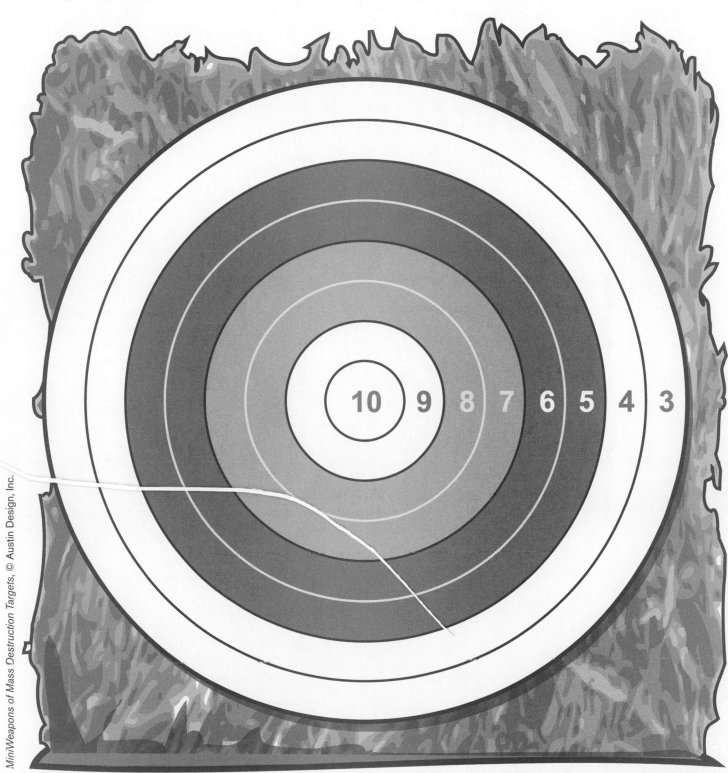

Competitor(s) _____   Date _____

MiniWeapon Used _____

ARCHERY TARGET RULES

Basic Rules ★

Alternating turns, each player fires three or six shots at the target from 5 to 12 feet away. When each player is finished shooting, competitors add up the score; play continues until everyone has had a turn. The shooter with the highest total score wins.

Marksman Rules ★★★

Alternating turns, each player fires three or six shots at the target from at least 10 feet away. The first marksman to score a cumulative 50 points or higher wins!

Competitor(s) _____   Date _____

QUINTAIN RULES

Basic Rules ★

Designed to assist knights with battle exercises, the quintain is the perfect target for practicing your Mini-Weapon skills. Hitting the helmet, shield, chest armor, or right arm with its mounted mace will knock this training dummy off guard. Alternate turns, with each player firing three or six shots at a distance greater than 6 feet. The player with the highest total hit count wins.

Marksman Rules ★ ★ ★

A true marksman will want to demonstrate his or her skill by aiming only at the helmet. One solid hit to this metal head wins!

MiniWeapons of Mass Destruction Targets, © Austin Design, Inc.

Competitor(s) _____   Date _____

MiniWeapon Used _____

QUINTAIN RULES

Basic Rules ★

Designed to assist knights with battle exercises, the quintain is the perfect target for practicing your Mini-Weapon skills. Hitting the helmet, shield, chest armor, or right arm with its mounted mace will knock this training dummy off guard. Alternate turns, with each player firing three or six shots at a distance greater than 6 feet. The player with the highest total hit count wins.

Marksman Rules ★★★

A true marksman will want to demonstrate his or her skill by aiming only at the helmet. One solid hit to this metal head wins!

Competitor(s) _____ Date _____

MiniWeapon Used _____

QUINTAIN RULES

Basic Rules ★

Designed to assist knights with battle exercises, the quintain is the perfect target for practicing your Mini-Weapon skills. Hitting the helmet, shield, chest armor, or right arm with its mounted mace will knock this training dummy off guard. Alternate turns, with each player firing three or six shots at a distance greater than 6 feet. The player with the highest total hit count wins.

Marksman Rules ★ ★ ★

A true marksman will want to demonstrate his or her skill by aiming only at the helmet. One solid hit to this metal head wins!

Competitor(s) _____   Date _____

MiniWeapon Used _____

QUINTAIN RULES

Basic Rules ★

Designed to assist knights with battle exercises, the quintain is the perfect target for practicing your Mini-Weapon skills. Hitting the helmet, shield, chest armor, or right arm with its mounted mace will knock this training dummy off guard. Alternate turns, with each player firing three or six shots at a distance greater than 6 feet. The player with the highest total hit count wins.

Marksman Rules ★★★

A true marksman will want to demonstrate his or her skill by aiming only at the helmet. One solid hit to this metal head wins!

Competitor(s) _____ Date _____

MiniWeapon Used _____

QUINTAIN RULES

Basic Rules ★

Designed to assist knights with battle exercises, the quintain is the perfect target for practicing your Mini-Weapon skills. Hitting the helmet, shield, chest armor, or right arm with its mounted mace will knock this training dummy off guard. Alternate turns, with each player firing three or six shots at a distance greater than 6 feet. The player with the highest total hit count wins.

Marksman Rules ★ ★ ★

A true marksman will want to demonstrate his or her skill by aiming only at the helmet. One solid hit to this metal head wins!

Competitor(s) _____ Date _____

MiniWeapon Used _____

CASTLE SIEGE RULES

Basic Rules ★

Alternating turns, each player fires three times at the castle's wooden gate from a distance of 8 or more feet. The shooter with most hits after three turns wins.

Marksman Rules ★★★

Stop the archers in the tower by hitting the windows! The first marksman to successfully hit one of the two windows wins. Because you are sharpshooters, move to a distance greater than 8 feet to play this game. Remember to alternate turns.

Competitor(s) _____ Date _____

MiniWeapon Used _____

CASTLE SIEGE RULES

Basic Rules ★

Alternating turns, each player fires three times at the castle's wooden gate from a distance of 8 or more feet. The shooter with most hits after three turns wins.

Marksman Rules ★★★

Stop the archers in the tower by hitting the windows! The first marksman to successfully hit one of the two windows wins. Because you are sharpshooters, move to a distance greater than 8 feet to play this game. Remember to alternate turns.

Competitor(s) _____   Date _____

MiniWeapon Used _____

CASTLE SIEGE RULES

Basic Rules ★

Alternating turns, each player fires three times at the castle's wooden gate from a distance of 8 or more feet. The shooter with most hits after three turns wins.

Marksman Rules ★ ★ ★

Stop the archers in the tower by hitting the windows! The first marksman to successfully hit one of the two windows wins. Because you are sharpshooters, move to a distance greater than 8 feet to play this game. Remember to alternate turns.

Competitor(s) _____ Date _____

MiniWeapon Used _____

CASTLE SIEGE RULES

Basic Rules ★

Alternating turns, each player fires three times at the castle's wooden gate from a distance of 8 or more feet. The shooter with most hits after three turns wins.

Marksman Rules ★ ★ ★

Stop the archers in the tower by hitting the windows! The first marksman to successfully hit one of the two windows wins. Because you are sharpshooters, move to a distance greater than 8 feet to play this game. Remember to alternate turns.

Competitor(s) _____   Date _____

MiniWeapon Used _____

CASTLE SIEGE RULES

Basic Rules ★

Alternating turns, each player fires three times at the castle's wooden gate from a distance of 8 or more feet. The shooter with most hits after three turns wins.

Marksman Rules ★★★

Stop the archers in the tower by hitting the windows! The first marksman to successfully hit one of the two windows wins. Because you are sharpshooters, move to a distance greater than 8 feet to play this game. Remember to alternate turns.

Competitor(s) _____ Date _____

MiniWeapon Used _____

APPLE RULES

Basic Rules ★

Alternating turns, each player fires three times at this apple from a distance of at least 8 feet. The shooter with most hits after three turns wins.

Marksman Rules ★ ★ ★

Pest control! The worm is worth 25 points and each fly is 50 points; the first player to score at least 100 points wins. Alternate turns, with each player firing three or six shots at the target from at least 10 feet away.

Competitor(s) _____   Date _____

MiniWeapon Used _____

APPLE RULES

Basic Rules ★

Alternating turns, each player fires three times at this apple from a distance of at least 8 feet. The shooter with most hits after three turns wins.

Marksman Rules ★★★

Pest control! The worm is worth 25 points and each fly is 50 points; the first player to score at least 100 points wins. Alternate turns, with each player firing three or six shots at the target from at least 10 feet away.

Competitor(s) _____   Date _____

MiniWeapon Used _____

APPLE RULES

Basic Rules ★

Alternating turns, each player fires three times at this apple from a distance of at least 8 feet. The shooter with most hits after three turns wins.

Marksman Rules ★★★

Pest control! The worm is worth 25 points and each fly is 50 points; the first player to score at least 100 points wins. Alternate turns, with each player firing three or six shots at the target from at least 10 feet away.

Competitor(s) _____ Date _____

MiniWeapon Used _____

APPLE RULES

Basic Rules ★

Alternating turns, each player fires three times at this apple from a distance of at least 8 feet. The shooter with most hits after three turns wins.

Marksman Rules ★★★

Pest control! The worm is worth 25 points and each fly is 50 points; the first player to score at least 100 points wins. Alternate turns, with each player firing three or six shots at the target from at least 10 feet away.

Competitor(s) _____ Date _____

MiniWeapon Used _____

APPLE RULES

Basic Rules ★

Alternating turns, each player fires three times at this apple from a distance of at least 8 feet. The shooter with most hits after three turns wins.

Marksman Rules ★★★

Pest control! The worm is worth 25 points and each fly is 50 points; the first player to score at least 100 points wins. Alternate turns, with each player firing three or six shots at the target from at least 10 feet away.

Competitor(s) _____ Date _____

MiniWeapon Used _____

SCARECROW RULES

Basic Rules ★

Alternating turns, each player fires at the scarecrow three times from a distance of 6 to 10 feet. The one with the most hits after three turns wins! Just avoid hitting the water bucket—do so and you'll lose a point.

Marksman Rules ★ ★ ★

Although it appears this scarecrow is just hanging out with his feathered friends, he's actually far more sinister and could attack at any moment! Stop him in his tracks by bouncing a shot off his head. The first marksman to wipe off his silly grin wins. Remember to alternate turns.

Competitor(s) _____ Date _____

MiniWeapon Used _____

SCARECROW RULES

Basic Rules ★

Alternating turns, each player fires at the scarecrow three times from a distance of 6 to 10 feet. The one with the most hits after three turns wins! Just avoid hitting the water bucket—do so and you'll lose a point.

Marksman Rules ★ ★ ★

Although it appears this scarecrow is just hanging out with his feathered friends, he's actually far more sinister and could attack at any moment! Stop him in his tracks by bouncing a shot off his head. The first marksman to wipe off his silly grin wins. Remember to alternate turns.

Competitor(s) _____ Date _____

MiniWeapon Used _____

SCARECROW RULES

Basic Rules ★

Alternating turns, each player fires at the scarecrow three times from a distance of 6 to 10 feet. The one with the most hits after three turns wins! Just avoid hitting the water bucket—do so and you'll lose a point.

Marksman Rules ★ ★ ★

Although it appears this scarecrow is just hanging out with his feathered friends, he's actually far more sinister and could attack at any moment! Stop him in his tracks by bouncing a shot off his head. The first marksman to wipe off his silly grin wins. Remember to alternate turns.

SCARECROW

Competitor(s) _____ Date _____

MiniWeapon Used _____

SCARECROW RULES

Basic Rules ★

Alternating turns, each player fires at the scarecrow three times from a distance of 6 to 10 feet. The one with the most hits after three turns wins! Just avoid hitting the water bucket—do so and you'll lose a point.

Marksman Rules ★ ★ ★

Although it appears this scarecrow is just hanging out with his feathered friends, he's actually far more sinister and could attack at any moment! Stop him in his tracks by bouncing a shot off his head. The first marksman to wipe off his silly grin wins. Remember to alternate turns.

Competitor(s) _____   Date _____

MiniWeapon Used _____

SCARECROW RULES

Basic Rules ★

Alternating turns, each player fires at the scarecrow three times from a distance of 6 to 10 feet. The one with the most hits after three turns wins! Just avoid hitting the water bucket—do so and you'll lose a point.

Marksman Rules ★ ★ ★

Although it appears this scarecrow is just hanging out with his feathered friends, he's actually far more sinister and could attack at any moment! Stop him in his tracks by bouncing a shot off his head. The first marksman to wipe off his silly grin wins. Remember to alternate turns.

Competitor(s) _____   Date _____

MiniWeapon Used _____

DRAGON SLAYER RULES

Basic Rules ★

Save the sheep! Any shot to this fire-breathing reptile should give you enough time to save the sheep and yourself. Just remember, he really doesn't like getting hit in the nose, belly, or wing! The first player to successfully hit one of the dragon's vulnerable areas wins!

Marksman Rules ★★★

It appears this dragon has scales as thick as armor, except on his tummy! This may be your only opportunity to save the village. The first player to hit his underbelly wins the game and the treasure!

MiniWeapons of Mass Destruction Targets, © Austin Design, Inc.

Competitor(s) _____ Date _____

MiniWeapon Used _____

DRAGON SLAYER RULES

Basic Rules ★

Save the sheep! Any shot to this fire-breathing reptile should give you enough time to save the sheep and yourself. Just remember, he really doesn't like getting hit in the nose, belly, or wing! The first player to successfully hit one of the dragon's vulnerable areas wins!

Marksman Rules ★★★

It appears this dragon has scales as thick as armor, except on his tummy! This may be your only opportunity to save the village. The first player to hit his underbelly wins the game and the treasure!

Competitor(s) _____   Date _____

MiniWeapon Used _____

DRAGON SLAYER RULES

Basic Rules ★

Save the sheep! Any shot to this fire-breathing reptile should give you enough time to save the sheep and yourself. Just remember, he really doesn't like getting hit in the nose, belly, or wing! The first player to successfully hit one of the dragon's vulnerable areas wins!

Marksman Rules ★ ★ ★

It appears this dragon has scales as thick as armor, except on his tummy! This may be your only opportunity to save the village. The first player to hit his underbelly wins the game and the treasure!

MiniWeapons of Mass Destruction Targets, © Austin Design, Inc.

Competitor(s) _____   Date _____

MiniWeapon Used _____

DRAGON SLAYER RULES

Basic Rules ★

Save the sheep! Any shot to this fire-breathing reptile should give you enough time to save the sheep and yourself. Just remember, he really doesn't like getting hit in the nose, belly, or wing! The first player to successfully hit one of the dragon's vulnerable areas wins!

Marksman Rules ★ ★ ★

It appears this dragon has scales as thick as armor, except on his tummy! This may be your only opportunity to save the village. The first player to hit his underbelly wins the game and the treasure!

Competitor(s) _____ Date _____

MiniWeapon Used _____

DRAGON SLAYER RULES

Basic Rules ★

Save the sheep! Any shot to this fire-breathing reptile should give you enough time to save the sheep and yourself. Just remember, he really doesn't like getting hit in the nose, belly, or wing! The first player to successfully hit one of the dragon's vulnerable areas wins!

Marksman Rules ★★★

It appears this dragon has scales as thick as armor, except on his tummy! This may be your only opportunity to save the village. The first player to hit his underbelly wins the game and the treasure!

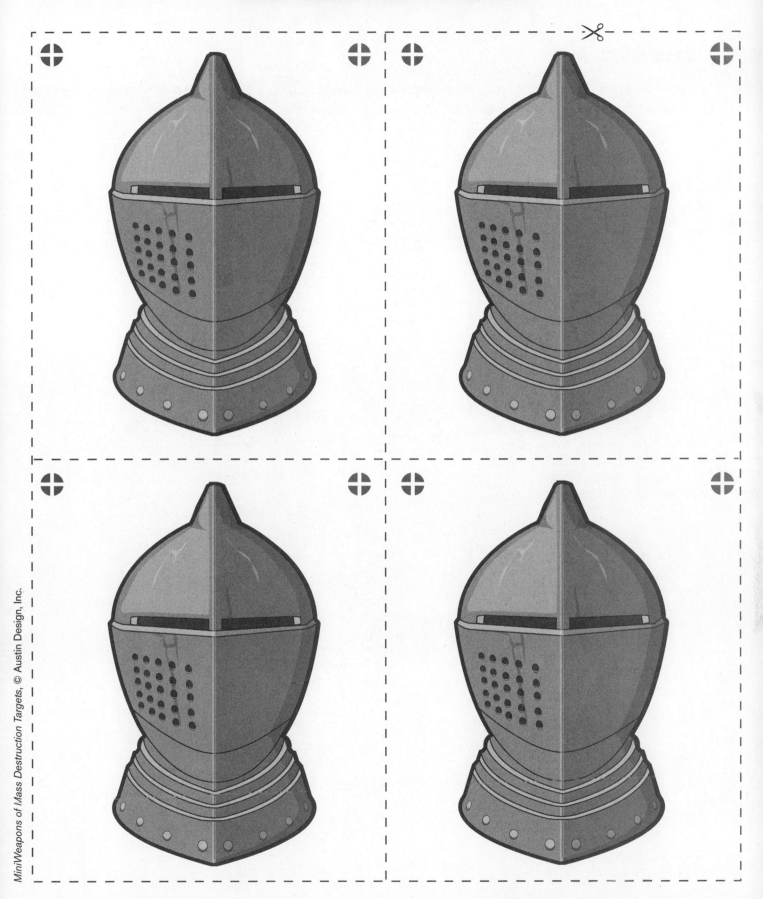

MINI DARK AGES TARGETS RULES

Basic Rules ★

Once cut apart, the Mini Dark Ages Targets are the perfect size for miniature arsenal testing on a wall or mounted on a tabletop. Use them for target practice or custom games.

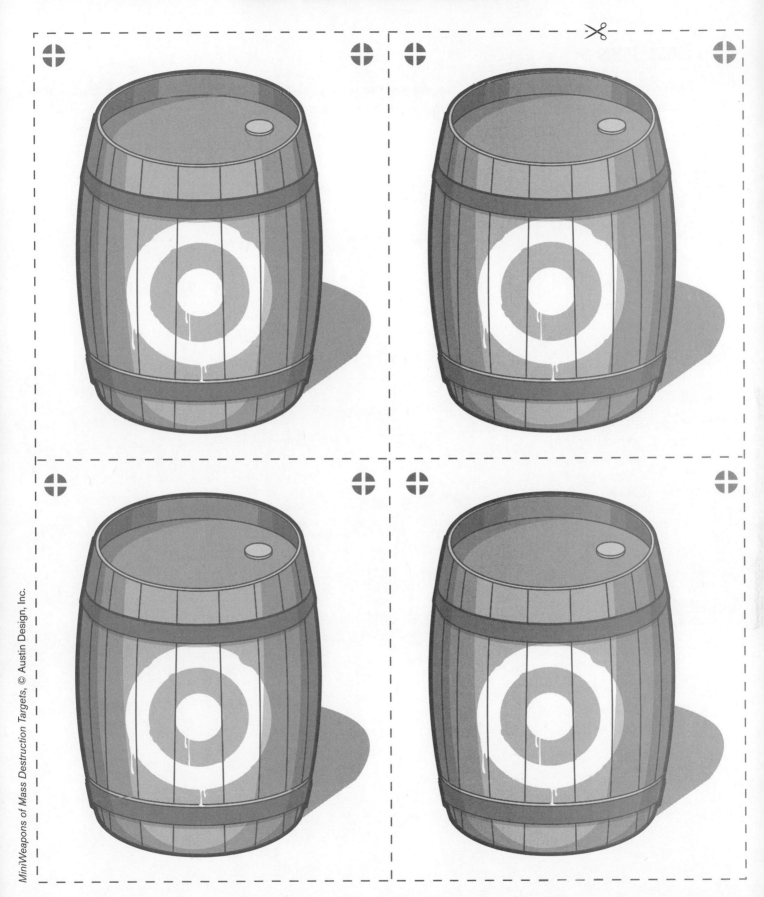

MINI DARK AGES TARGETS RULES

Basic Rules ★

Once cut apart, the Mini Dark Ages Targets are the perfect size for miniature arsenal testing on a wall or mounted on a tabletop. Use them for target practice or custom games.

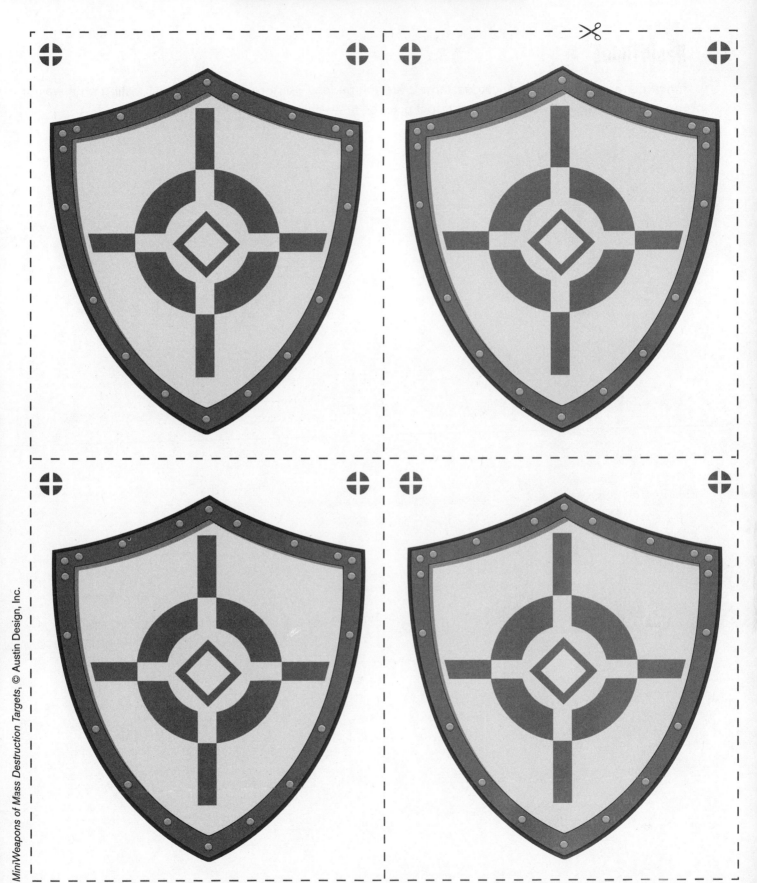

MINI DARK AGES TARGETS RULES

Basic Rules ★

Once cut apart, the Mini Dark Ages Targets are the perfect size for miniature arsenal testing on a wall or mounted on a tabletop. Use them for target practice or custom games.

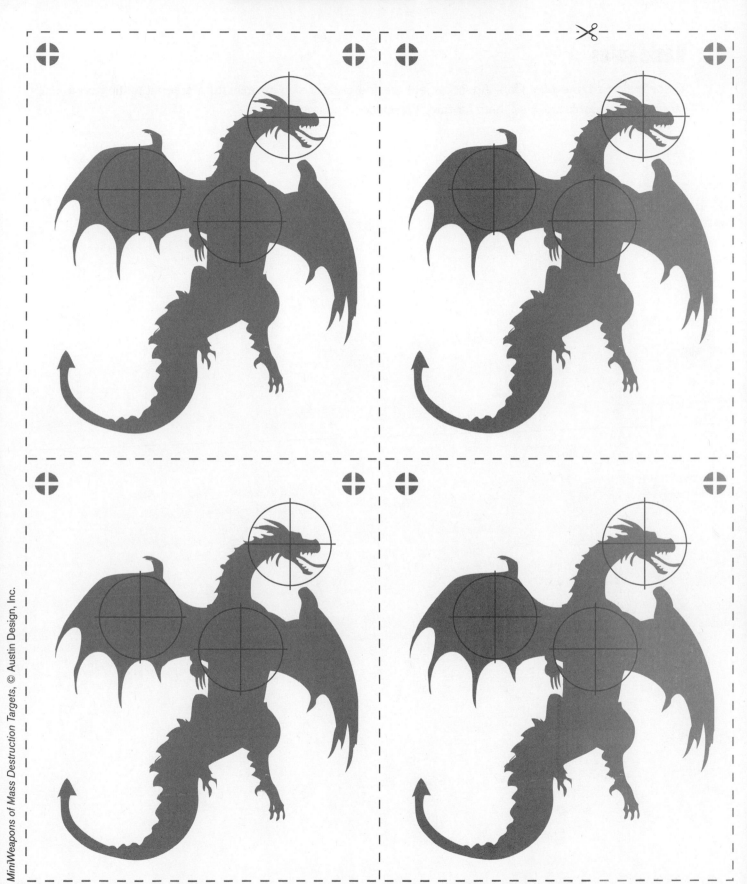

MINI DARK AGES TARGETS RULES

Basic Rules ★

Once cut apart, the Mini Dark Ages Targets are the perfect size for miniature arsenal testing on a wall or mounted on a tabletop. Use them for target practice or custom games.

For more information and more free downloadable targets, please visit:

MINIWEAPONSBOOK.COM

DON'T FORGET TO JOIN THE MINIWEAPONS ARMY ON FACEBOOK:

MiniWeapons of Mass Destruction: Homemade Weapons Page